3D PRINTERS
A Beginner's Guide

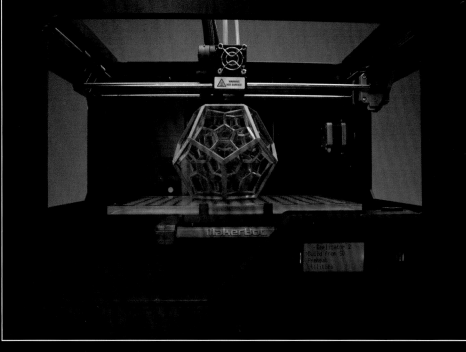

OLIVER BOTHMANN

3D PRINTERS

A Beginner's Guide

FOX CHAPEL
PUBLISHING

© 2015 by Oliver Bothmann and Fox Chapel Publishing Company, Inc., East Petersburg, PA.

First published in the United Kingdom by Special Interest Model Books, 2014.
First published in North America in 2015 by Fox Chapel Publishing, 1970 Broad Street, East Petersburg, PA 17520.

ISBN 978-1-56523-871-8

Library of Congress Cataloging-in-Publication Data

Bothmann, Oliver, author.
  3D printers / Oliver Bothmann.
    pages cm
  Summary: "Provides a detailed explanation of the basics of purchasing and using 3D printers for total beginners."--
Provided by publisher.
  Includes bibliographical references and index.
  ISBN 978-1-56523-871-8
  1. Three-dimensional printing. I. Title. II. Title: Three D printers.
  TS171.95.B68 2015
  621.9'88--dc23

                    2015000629

To learn more about the other great books from Fox Chapel Publishing, or to find a retailer near you, call toll-free 800-457-9112 or visit us at *www.FoxChapelPublishing.com*.

Note to Authors: We are always looking for talented authors to write new books. Please send a brief letter describing your idea to Acquisition Editor, 1970 Broad Street, East Petersburg, PA 17520.

Printed in China
First printing

Contents

Gallery

Over the past years, the technology of 3D printing has grown by leaps and bounds. Many companies and individual artists have been working to push the rapidly expanding limits of the field and create unique, often functional, and always impressive items with their printers of all sizes. Featured in this gallery are just a handful of inspiring examples of stunning pieces.

The MGX Quin lamp creates its visually striking effect through its 3D-printed shade. (Image courtesy Bathsheba Grossman, www.bathsheba.com.)

One of the largest models of 3D printer available, the BigRep printer, created this side table in one piece. See chapter 3 for more information about different printers. (Image courtesy BigRep, www.bigrep.com)

This flexible dress, the Kinematics Dress, is made from interlocking, articulated modules and is printed as one folded piece. (Image courtesy Nervous System, http://n-e-r-v-o-u-s.com; photo by Steve Marsel)

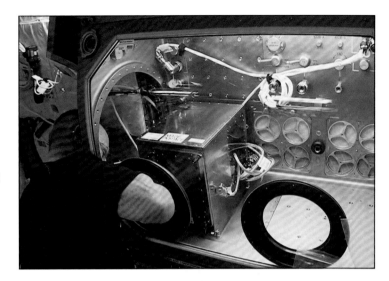

The 3D printer in the International Space Station was shipped into space and first used in late 2014. In this photo, Commander Barry "Butch" Wilmore is setting up the machine in its new home. (Image courtesy NASA-TV)

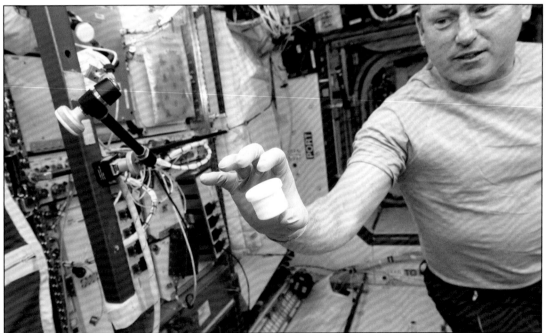

International Space Station Commander Barry "Butch" Wilmore poses with a science sample container, one of the first items to be printed in space. For more 3D printing from NASA, see page 55. (Image courtesy NASA)

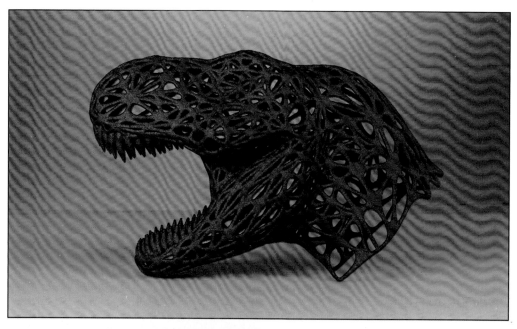

The 3D-REX sculpture showcases an interesting capability of 3D printing, a webbing-like, hollow build effect. (Image courtesy Octavio Asensio, www.octavioasensio.com)

This hollow piggy bank, called Oinky, was printed from a strong, flexible nylon material. (Image courtesy Octavio Asensio, www.octavioasensio.com)

Interlacing filaments create a beautiful effect in the Mobius Nautilus sculpture. (Image design and photography by Joaquin Baldwin, www.shapeways.com/shops/joabaldwin)

Believe it or not, this vase was 3D-printed in sugar. (Image courtesy 3D Systems, www.3dsystems.com)

Dyed sugar cubes make 3D printing both beautiful and edible. (Image courtesy 3D Systems, www.3dsystems.com)

Even food, like sugar and chocolate, can be 3D-printed with special printers. These 3D-printed sugar cubes are more like mini sugar sculptures. (Image courtesy 3D Systems, www.3dsystems.com)

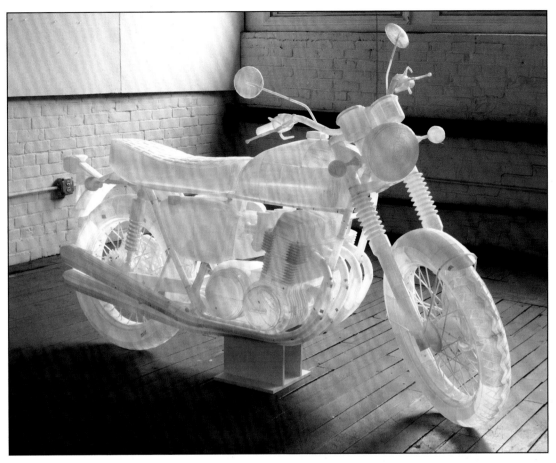

3D printing can produce amazing results when many individually printed pieces are combined. The life-sized motorcycle replica above contains all the moving parts except for a motor—it just doesn't actually drive! (Image courtesy Jonathan Brand, www.jonathanbrand.com)

Every piece of the motorcycle model was 3D-printed. (Image courtesy Jonathan Brand, www.jonathanbrand.com)

The Casa Fortunata, a replica of a real home, was printed in exquisite detail and painted. (Image courtesy White Clouds, www.whiteclouds.com)

Individual windows, bricks, roof textures, and many more details were included in this house model. (Image courtesy White Clouds, www.whiteclouds.com)

The Biomorphic iPhone 6 Cover is stylish, functional, and inexpensive, showing just how far 3D-printed items have come in terms of accessibility and utility. For more ideas for useful objects to print, see chapter 12. (Image courtesy Sam Abbott, www.samabbott.com)

This vase sculpture is an example of the kind of smooth, flowing lines that can be achieved with masterful 3D printing. (Image courtesy Noah Hornberger, www.etsy.com/shop/MeshCloud)

Introduction

Stories about 3D printing appear almost everywhere in the media today. They are often about the danger of weapons being printed at home or things like "You could print yourself a pizza!" This may be interesting for readers of newspapers or TV channel viewers, but it is not the most important aspect of 3D printing.

This new technology is a revolution compared with production processes as we know them. In the future there may even be 3D printing shops, similar to a copy shop. You can already send a digital file of a part, which you have created, to a company that will print it in plastic, steel, or brass, or even in titanium, gold, or ceramic.

We can also use this 3D printing technique at home, although only printing in thermoplastic is possible at the moment. However, even with this limited material, it is possible to print almost anything. You can print decorative objects and special toys. It is also possible to print spare parts for nearly everything, such as an old radio or your car. You can print personalized objects, like key rings or bracelets, with your name or the name of your friend on them. As you learn more about the technology, you will find there are a lot more things you can do with a 3D printer at home.

This book will guide you through your first steps in 3D printing. It will show you what is possible and what is not. You will be shown what you need to do to make your first 3D-printed part and what you have to learn to become a 3D printing professional.

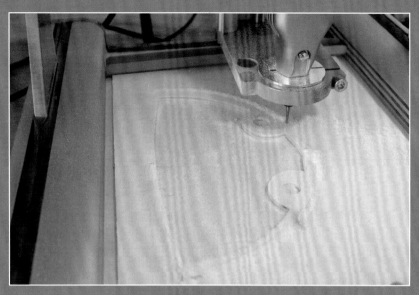

02.01 During subtractive production processes, like this milling process, material is removed to produce the desired shape.

02.02 During additive production processes, like this 3D printing process, material is built up, minimizing waste.

What is 3D printing?

If you have ever watched Star Trek, you will know about the "Replicator." It is the machine that produces everything that the crew wants just by speaking to it. Certainly, 3D printing so far is not the same thing, although the idea behind it seems similar. The US 3D printer company Makerbot actually named their 3D printers "Replicator".

In science fiction programs, these replicating machines can build anything. Today it is only possible to produce things from of a limited range of materials, although in such a way that it is still revolutionary. Almost every machine tool works in a so-called subtractive way, which means that a lathe or a milling machine removes material from a larger block to make the part. 3D printers work in a totally different way. They build the part, layer by layer, in an additive way until the part is complete. **(Illus. 02.01), (Illus. 02.02)**

This additive process has some advantages over the subtractive process where a lot of material is "lost" during production. Of course, this material can be recycled, but, for the component manufacturer, it is gone; they have paid for a lot of material which will never be used. In the additive production method, only the material needed will be printed onto the part, which means it will use considerably less material to make it.

In an additive building process, you can build parts that are not possible to make in a subtractive process. For example, it is possible to make parts in on piece that include undercuts! With subtractive machining, many parts were not possible to make and had to be

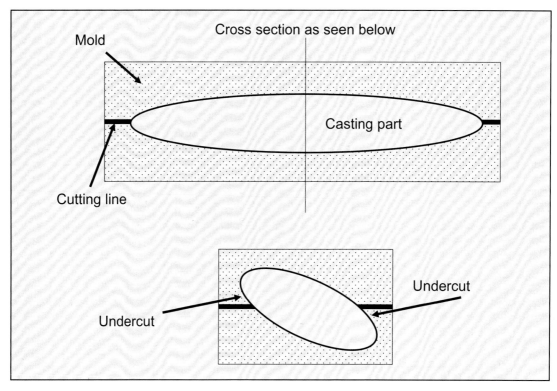

Mold

Cross section as seen below

Casting part

Cutting line

Undercut

Undercut

02.03 A drawing of a mold with undercuts. This molding would be a problem with normal molding but not with 3D printing.

made as multiple parts and fixed together after machining. **(Illus. 02.03)**

3D printing, often named "rapid prototyping," is not a single process. There are several technologies that are pooled together and called 3D printing. In this overview, I will describe some of the most common 3D printing techniques.

One of the oldest techniques in this field is stereolithography (STL). Here, a photosensitive fluid, normally a kind of resin, is used. At the start of the process, a platform is lying just under the surface of a special fluid so that there is a thin layer of fluid on the platform. An ultraviolet light (or laser) is moved over the surface and where the light hits it, the resin hardens. After the first layer is finished, the platform is lowered in the fluid by the thickness of the next layer. The fluid runs over the first layer and the light hardens the next layer. After the process is finished, the platform rises out of the fluid and the printed object can be taken from the machine and cleaned. There are some machines of this type suitable for home users, but they are very expensive.

There are two problems with stereolithography: working with the resin (which is not very healthy) and the disposal of the unused material. This makes it an

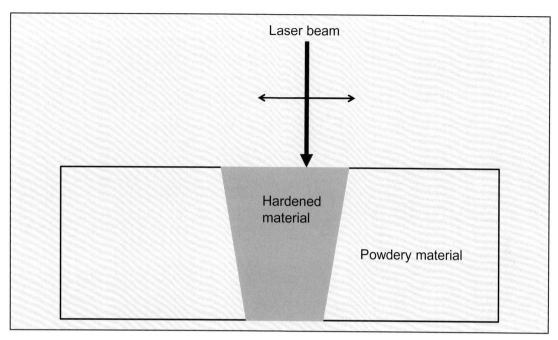

Laser beam

Hardened material

Powdery material

02.04 A schematic drawing of the SLS-technique.

impracticable kind of 3D printing for home users.

Another 3D printing technology is selective laser sintering (SLS), which is the main technology that is used in industry and by service providers who do 3D printing. The process uses a laser to melt (sinter) powdery material (plastic, metal, or even ceramic) together so that it makes a solid part. Sometimes, especially with plastics, the powder must be specially treated so the laser can melt the material together.

This process is very similar to stereolithography. A thin layer of powder is lying on a platform with the laser moving over it, and at the areas where the part should be formed, it will melt the powder together. Then the platform lowers by the thickness of the next layer and another layer of powder is

added over the first layer. The laser sinters the powder and the process is repeated until the part is completed. The platform will then rise so the part can be removed and cleaned of unused powder. The unused powder can be recycled back into the production process.

SLS is a technology that is too expensive (around tens of thousands of dollars) for the home user, is very difficult to use, and requires (expensive) training. If you order 3D prints from a service provider, they will normally be done using this technique. **(Illus. 02.04), (Illus. 02.05)**

The most common 3D printing technique for home users is fused deposition modeling (FDM). This process employs a thermoplastic, normally in the form of a filament on a spool. The thermoplastic is heated and then ejected as a molten thread from a heated

02.05 The Plastic-Laser-Sinter-System Eosint P 395 from EOS *(Picture: EOS).*

nozzle. The fine, hot plastic thread will be "printed" in layers on a bedplate. Due to the high temperature of the nozzle, the plastic threads will melt together and become a stable part.

One of the advantages of FDM is that there is no toxic waste. If you use bio-plastic like PLA (I will explain materials later in this book), there is not even an objectionable odor. The usage of this technique is also very simple and nearly all printers designed for home use work in this way.

The main disadvantage is that the printed objects normally have "steps" resulting from the small plastic threads that are quite visible. With some experience, and trial and error, you can reduce the steps to

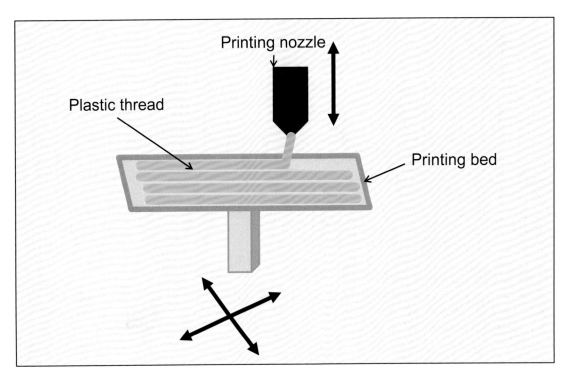

02.06 A schematic drawing of the FDM-technique.

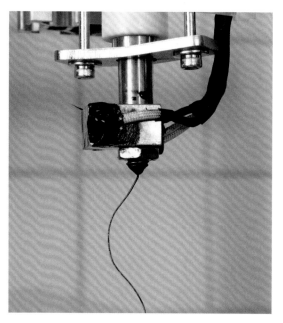

02.07 In this picture you can see how the thread of melted plastic leaves the printing nozzle.

02.08 It is hard to believe, but this bust is not made out of marble. It's made of paper. *(Picture: Mcor).*

a minimum but not stop them completely. If you want a really smooth surface, you will have to sand it smooth or use some filler. There are some techniques using solvents to make the surface smooth, but all of them are more or less unhealthy, and so I will not give any advice for doing this. **(Illus. 02.06), (Illus. 02.07)**

There is a form of 3D printing that builds parts out of layers of foil (plastic) or even normal paper. The material will be laminated, which is the reason it is called "laminated object modeling" or LOM. The material will be glued one layer over the next and then cut to the required shape with a knife or a laser.

One of the leading companies using a special form of this technique is the

Irish enterprise Mcor technologies (www.mcortechnologies.com). The printer works with normal writing paper that is then glued together and cut with a Tungsten Carbide blade. The paper can be printed with an inkjet printer prior to lamination for making colored printed parts. For part samples or educational purposes, this is a great advantage. Due to the use of paper and water based glue, this technology is also very eco-friendly.

Unfortunately, these printers are unlikely to be used in the home; they are expensive and only available at service providers. Perhaps there will be a printer using this technique for home users in the near future. **(Illus. 02.08)**

Probably one of the best-known 3D printers for home use is the Replicator, made by the US company Makerbot. *(Picture: Makerbot®). (see page 26)*

3

3D printers

The 3D printer market, especially for the home user, is rapidly expanding. Nearly every day there is a new printer announced, and new companies are springing up everywhere. Most of the printers are based on the FDM technology, because most of the patents for this technique expired some time ago.

A lot of these companies are based in the USA, but it is very interesting that, unlike in the normal office printer business, there are a lot of small and innovative firms developing and producing different types of 3D printers. Great Britain, the Netherlands, Germany, Israel, the Czech Republic, and some other countries are the home of such start-up companies.

Due to the fact that nearly all of the printers are based on the FDM technology, their main principles are similar. A plastic filament (mostly with a diameter of 1.75 to 3 mm) is melted in a hot end and pushed through a nozzle with an appreciably thinner diameter of typically 0.5 mm. These thin threads create the printed parts. Although these machines all use similar principles, the printers themselves differ in many ways.

In the early years of home 3D printing, printers were mostly self-built machines, and you built them from component parts. The so-called Rep-Rap-community was the starting point for most of the companies that now produce printers for home users today. A lot of knowledge is required to self-build a 3D printer, not only mechanical knowledge, but also knowledge of electronics and computer programming. Because of the knowledge required, only a very small number of enthusiasts were able to build a 3D printer of their own.

The rise of 3D printers during the last few years has been possible because of the complete kits and ready-to-use printers that were available. Even people that could not or were not willing to build a printer of their own were willing to assemble a kit or buy a working 3D printer.

The main difference in the printers offered is that they are sold as a kit or ready-to-use. Often the kits are cheaper than the printers that come ready-to-use, but this is not the only difference. If you buy a kit, you should have a look at the quality of the components that are used. Sometimes, the kits are made out of components that are not ideal for building a printer. Some companies use threaded rods for the main frame of the printer. It is possible to do so, but the fine

adjustment of a printer built in this way is complex and not practical for someone that wants an easy-to-use printer. The best way is to use special connecting sections (made from extruded aluminum) because they are stable and easy to adjust. **(Illus. 03.01)**

If the kit is made from quality components and you want to do the assembly on your own, it is a good idea to do so. You will know nearly every single part of your printer (and how to mount and dismount it), and, during the fine tuning, it will be much easier for you to find the solution to any problems that arise. So if you do not have two left feet, building a 3D printer from a kit may be the best option for you. The kits are usually easy to assemble using a few tools in a short time. **(Illus. 03.02), (Illus. 03.03)**

Where the first printers were mostly kits, there are now more and more ready-to-use

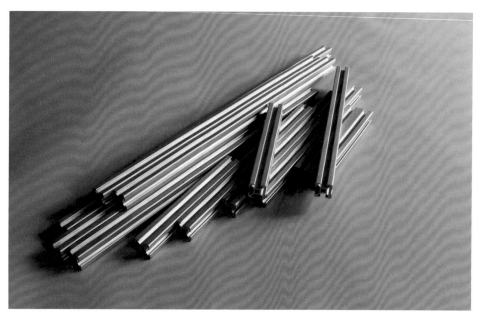

03.01 The Multirap by Multec is built out of high-quality aluminum profiles.

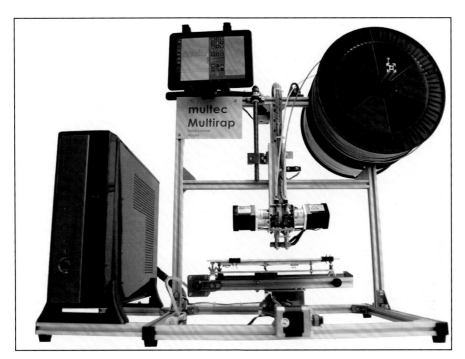

03.03 The Multirap from the German company Multec is supplied as a kit. *(Picture: Multec)*

03.02 Assembling the printer is very easy.

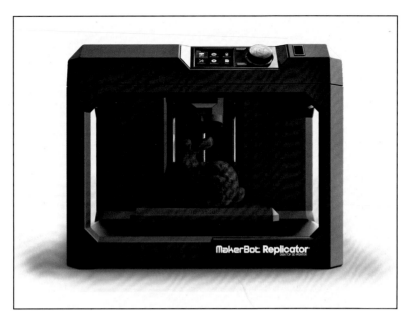

03.04 Probably one of the best-known 3D printers for home use is the Replicator, made by the US company Makerbot. *(Picture: Makerbot®)*.

03.05 The new CEL-Robox-Printer from the British company CEL-UK. *(www.cel-robox.com, picture: CEL-UK)*.

printers, probably because a lot of interested buyers are unwilling to build the machine on their own and because a ready-to-use printer normally does not need as much support as a kit-built printer.

There are an unimaginable number of ready-to-use printers: from simple machines with poor quality to high-end printers that will cost you a lot of money. To find the right printer for you is not easy, and, because of the fast-growing variety, it is not getting any easier.

The first advice I would like to give you is to only buy a printer that you have seen and/or that you know works well. To buy a cheap, nameless printer, maybe direct from the Far East, could be the first step to a fiasco. Of course, a lot of ready-to-use printers are produced in China, but I think it is important to use a distributor with a contact person in your home country. Small (or even big) problems will be easier to solve and your warranty should be valid.

Buying a printer that is totally enclosed or has a relatively open construction is not just a visual or design question. A totally enclosed printer will normally be more silent and, if you use plastics like ABS, they will not emit as much odor as open printers. Another point is that FDM printers are a little bit more susceptible to changes in room temperature, because the printed material can cool down too fast if the room is very cold. Therefore, a closed printer will be more resistant to temperature change. Some expensive printers have a heated chamber so that you can print in a cold environment, but for normal use this is not important.

03.06 Ultimaker printers are very well known (shown here is the Ultimaker2) and are made by the Dutch company Ultimaker. *(Picture: Ultimaker).*

03.07 US company 3D Systems creates a Cube machine for home use. *(Picture: 3D Systems).*

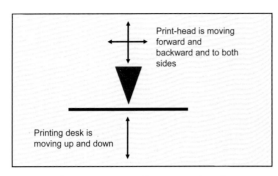

03.08 In this construction the printing head is moving forward and backward (Y-axis) and to both sides (X-axis). The printing desk is moving up and down (Z-axis).

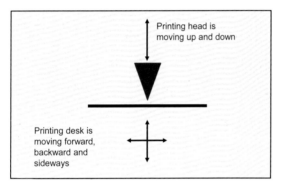

03.09 In this construction the printing head is moving up and down (Z-axis) and the printing desk forward, backward and sideways (X and Y-axis).

3D printers with a closed casing can make it hard to reach some parts of the mechanics if you need to repair or convert something. So if you would like to experiment with the mechanics of your printer, or you would like to upgrade some parts, it would be better to have an open frame printer.

A closed printer would, of course, be a better choice if you need to use your printer on your desktop like a normal printer. The designs of these printers vary from very

conservative black boxes to spacy-looking machines, but the look is not as important as the overall quality of the printer. **(Illus. 03.04), (Illus. 03.05), (Illus. 03.06), (Illus. 03.07)**

The main technical difference in the construction of FDM printers is the way in which the axes are moved. There are three axes that have to be moved. Moving up and down, which defines the thickness of the layers you will print, is called the Z-axis. Moving sideways to the left and right is normally called the X-axis, and the one moving backwards and forwards is the Y-axis. This is the normally used notation, although sometimes some manufacturers interchange the X and Y-axes.

In general, there are two ways in which the printing will be done. The first is that the bed on which the part will be printed is moved down (Z-axis) and the print head (in which the plastic is melted) will move over the desk sideways, backwards, and forwards (X and Y-axis). The other possible way is that the bed is moved sideways, backwards, and forwards (X and Y-axis) and the print head is moved up (Z-axis). It does not matter which way a printer moves if it is well adjusted. **(Illus. 03.08), (Illus. 03.09)**

An important factor in getting good results from your printer will be the method used to transmit the stepper motor movements to the bed and the print head. Possible methods are (with no claims of being complete) threaded spindles, toothed rods, and drive belts – and usually a combination of them. Often the X and Y-axis are driven with timing belts and the Z-axis with threaded rod, and this usually works well. Make sure that the Z-drive spindle has a good

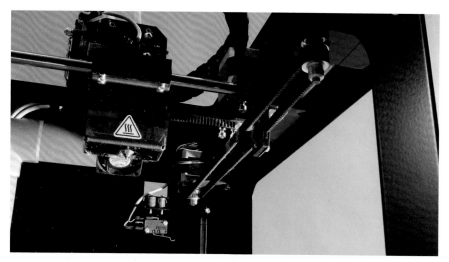

03.10 Toothed belts are often used for driving the moveable parts of 3D printers.

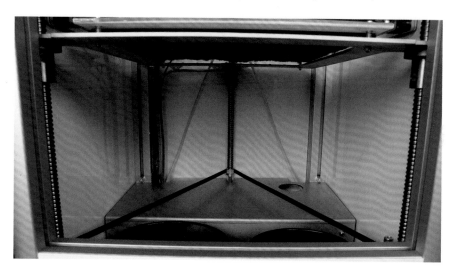

03.11 Exact movement is possible using threaded spindles like the three shafts in this printer.

quality trapezoidal thread, because this kind of spindle combines accuracy with good movement. Sometimes, mostly in self-built printers, threaded rods are used, but this is inadvisable! Threaded rods are made for attaching something to another item, but are not designed for moving. They may be cheap, but they are not as good as a threaded spindle that is designed for transmitting movement. **(Illus. 03.10), (Illus. 03.11)**

Another very important constructional feature is the solidity of the basic machine's

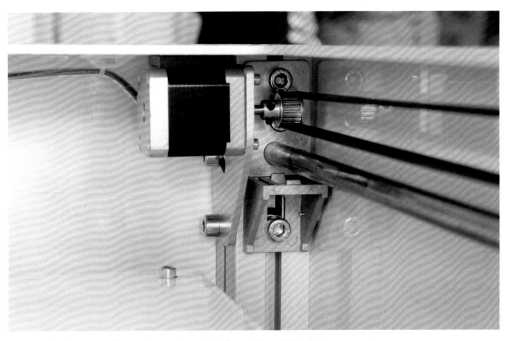
03.12 Aluminum profiles make a very stable frame for a 3D printer.

frame. Because all of the mechanical components are attached to the frame, it should be as rigid as possible. The stepper motors and mechanical drives will sometimes move jerkily, and, if the frame is not stable enough, these jerky movements will be transmitted through the frame and induce the whole printer to vibrate, causing inaccurate printing. (**Illus. 03.12**)

If you buy a ready-made 3D printer, you should make sure that it is really complete, as not all of them come with all the items you need! Ready-to-use printers normally do; you will get the printer itself, the electronic controls, and the software for driving the printer. If you buy a kit, you should look at what parts are included. Some kits will have only the mechanical components, while some will have all of the parts required to build the printer. If you are unsure, contact the manufacturer or dealer so you can order all of the parts that you need at one time. If you have to wait for an additional delivery, the waiting time will be very frustrating.

As well as the finished printer or kit, you will need some material for your printing. Some printers will have materials included and some will not, so do check this! The first thing needed is the filament, which is the basic material for printing. If you need any more filament or another color of filament, you can order it later.

Some printers need material for making the printouts adhere to the printing bed. This may be some kind of special glue or tape which you will have to put on the printing

03.13 As well as the printer, you need other items like filament and, often, adhesive tapes or glues.

bed. This may be delivered with your printer, but, if not, order it direct. Otherwise you will be waiting a while to print your first test successfully. (**Illus. 03.13**)

There are, as I will explain later, different plastics with varying characteristics that you could use for FDM 3D printing. Due to these characteristics, the printers may need special abilities to handle them. As an example, if you want to print ABS, which has a higher melting point than PLA, you will need a printer that can reach this higher temperature; not all printers are equipped for this.

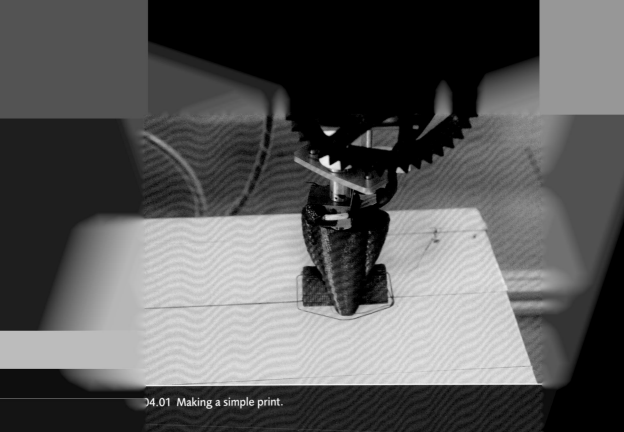

04.01 Making a simple print.

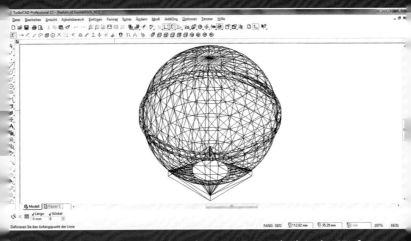

04.02 In this screenshot you see how the structures are built in the STL format.
Every surface is pieced together out of lots of little triangles.

The first print

You have your brand new 3D printer, but how and what should (and could) you print with it? In this chapter, I will show you how to make your first prints. I will use software that I know well and that I have used for some time. In the following chapters, I will also show you some other programs I use, so you should check them out and choose the software you like the best. **(Illus. 04.01)**

The first step for a print would be to get a file of the part you would like to print. First, you need a 3D drawing of the part. There are several data formats for such 3D drawings, but most 3D printers (or, more precisely, their driving programs) work with the common STL format. STL stands for Surface Tessellation Language; sometimes it will be translated as Stereolithography, and

a malapropism of the abbreviation is "stupid triangles, lots of them". Even if this is meant as a joke, it describes the speciality of STL well: the format defines a 3D design with a lot of triangles. The more triangles that are used, the better and smoother the print will be, because the triangles will be smaller. You can normally choose the number of triangles in the options part of your software. **(Illus. 04.02)**

There are some other file formats which can be used for producing the g-code (we will go into g-code later) that drives the printer, but the STL format is the most common data exchange format that will be outputted by most construction programs.

For your first prints, I would recommend that you use a ready-to-use STL file, which has ideally been printed by other users, so

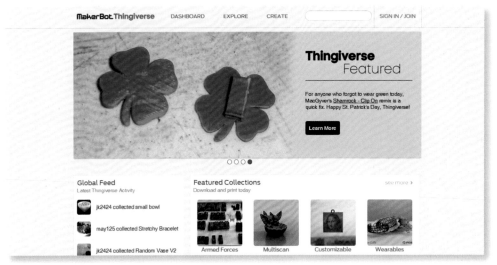

04.03 The home page at www.thingiverse.com; here you will find a tremendous amount of interesting objects to print. *(Picture: Thingiverse).*

that you know that the file is printable. If you use a self-made construction, there could be errors in your file, and the printer won't work properly, and you may wrongly blame the printer for that.

Most printer manufacturers supply a selection of test files for your first prints. Often they are in STL format and sometimes in ready-to-use g-code format, which the printer uses. Use these files for your first tests, because they are known to work properly.

If you have no test file from your printer manufacturer, you can use a file from one of the internet websites where files for 3D printers are offered free of charge. There are several of these websites, but perhaps the best two, which have a very wide range of construction files, are www.thingiverse.com and www.youmagine.com. I will expand on the websites and their philosophies later in this book. **(Illus. 04.03)**

Choose a design that is not too complex or too large. Normally there is some information included with these files, which you should read carefully. Often, in the information given, it will say if the object has already been printed or if it is only a virtual construction. Later, when you know your printer is working, you can try out some of these files to check if the data is working and give feedback on the website for other users. At first, only use tried and tested files.

When you have the file that you want to use for your first print, you will have to prepare it for printing. Sometimes printer manufacturers promote their printers with words like "Put an SD-card with the object you need in the slot of the printer and a little later you will have the finished part in your hands!" This may work sometimes, but normally it doesn't.

A 3D printer is not like a normal desktop printer for letters. It is a machine that needs,

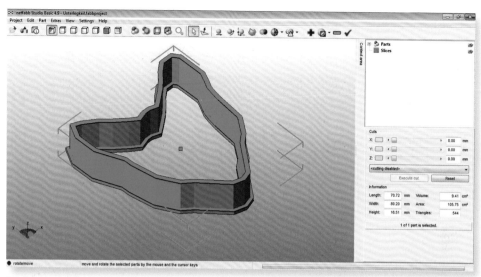

04.04 Netfabb is a very powerful tool for preparing your files for printing.

at least nowadays, a lot more knowledge of the process of preparing and producing the print.

There are some steps in the preparation of the files you send to the printer to make it do what you want.

First, you should check the file to see if it is printable. Sometimes STL files have errors in them, which means that there are gaps in the surface or there are other problems that will make the print flawed or impossible to print. Fortunately, there are programs that can detect and correct such errors automatically and do much more. The software I use is netfabb studio Basic, which is sufficient for normal use and is free as well.

Netfabb studio Basic is a very powerful tool with which you are able not only to check and repair files, but to scale them and much more. I will show you the important features of this program, and, if you work with it, you will find many more possibilities that may be helpful for you.

You can start netfabb either by starting the program itself or by clicking on an STL file if you have associated STL with netfabb. In this case the program will open the STL file on the netfabb screen; otherwise you will have to load the file up manually. **(Illus 04.04)**

You can now look at the part you want to print from every side and check if it is correct. The most important information will be shown on the bottom right-hand edge of the screen. If there is a warning sign with an exclamation mark in it, there will be some problems printing the file. Possibly the file has holes or incorrectly calculated triangles in it. But don't panic: netfabb will help you to solve these problems easily. **(Illus. 04.05)**

The next step is to analyze the problems. You should click under "Extras">"New Analysis" onto "Standard Analysis". You will then see in a window at the right-hand side which errors are in your file, e.g. if there are holes in it or incorrectly oriented triangles.

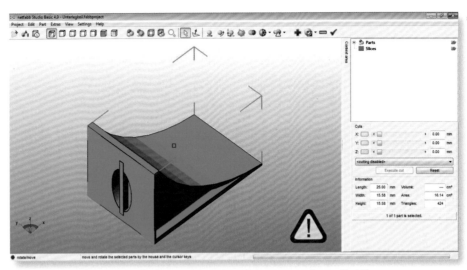

04.05 The "traffic sign" at the lower right edge of the screen shows that there are some errors in this file.

Even if the detail of your first print is a little bit confusing, it will become interesting as you get more printing experience. **(Illus. 04.06)**

For your first test prints, the easiest way to deal with these errors will be to let netfabb repair them automatically. Click under "Extras" on "Repair part" or on the red cross on the toolbar at the top of the screen. The display will change and the errors will be shown on your part drawing. At the

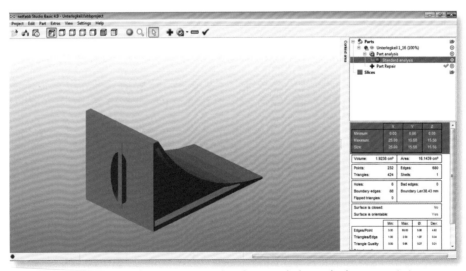

04.06 After analyzing the file, you can see that there are holes and other errors in it.

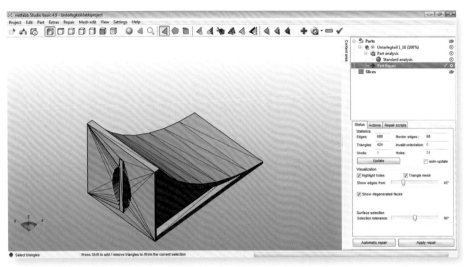

04.07 After a click on the repair-button, you will see this screen.

right-hand side you will be shown how many errors there are in the file.

Click on the button "Automatic repair" and in the dialogue on "Default repair" and "Execute." Netfabb will now correct the defects and, if you click on "Update" on the right-hand side (or activate "Auto-Update"), you will see that the errors have been removed.

Now click onto "Apply repair" and the dialogue will ask if the old part should be removed; click "Yes" and you will normally now have a file that is free of any errors. The

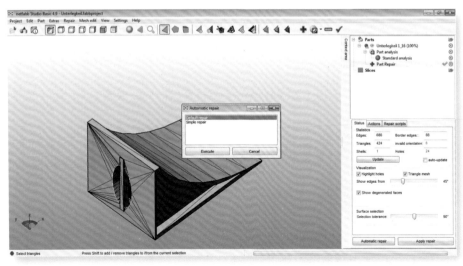

04.08 Click on "Automatic Repair" and execute this.

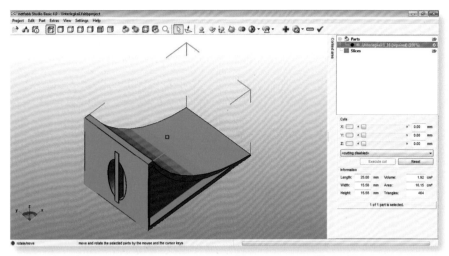

04.09 After clicking on "Apply Repair," you will have a correct file and can save it.

file is marked by an appendix "(repaired)" so you can identify it easily. **(Illus. 04.07)**, **(Illus. 04.08)**, **(Illus. 04.09)**, **(Illus. 04.10)**

This is your first step to getting a correct file completed. The next step will be to orient it correctly. Sometimes, during the construction process, the part will not be oriented correctly. For example, the print may want to print lower in the Z-axis (you will see a negative figure for this position) than is possible. This could be a real physical collision that should be prevented by the printer's safety limit switches, but the print will not be correct.

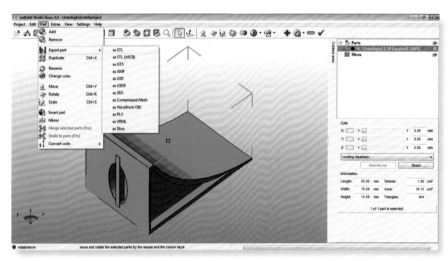

04.10 You must export the file to an STL file.

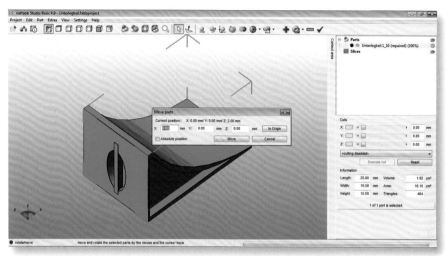

04.11 Unfortunately, the part is not always oriented correctly in the three-dimensional space, but with netfabb, it is easy to bring it to the origin with a single click.

Before printing, you will need to orient the part. This is easy to do; click under "Part" onto "Move" and in the following dialogue first click "to Origin" and then on "Move," which makes the object move to a neutral point in all three axes. (**Illus. 04.11**)

After this, you have to save the object as an STL file. Go to "Part" and then click on "Export part" > "as STL" and name it what you like. If you click "Save" there may be a warning that there are still some errors in the file, but if you then click on "repair," these

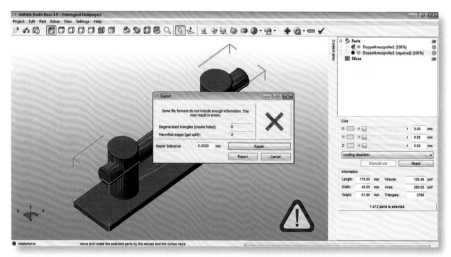

04.12 Sometimes more errors occur during the export. You can debug the errors with just one click.

will be corrected automatically and they can be saved (with the button "Export"). The file should then be flawless. (**Illus. 04.12**)

These repair options are not the only modifications netfabb can do for you. Netfabb has a very nice scaling tool with which you can easily change the size of the part you want to print. If you go to "Part" and click onto "Scale" you can enter the scale that you want your part changed to (e.g. halved or doubled in size) and the program will calculate and change it for you. With netfabb you have a very powerful tool for making files error-free and ready to print. (**Illus. 04.13**)

The next step is to generate the information the printer needs to print the part. The printer has to be told where to print the material using g-code. G-code is a fairly standardized programming language that is used for driving CNC (Computerized Numerical Control) machines and also 3D printers. Generating this code will be done by special programs that are normally supplied with your printer. Some programs are different in some details, but in the basic mode of operation they are comparable.

Because an FDM home 3D printer prints the part in layers, the part has to be divided into these layers. This process is called slicing, because the part will be cut, like bread,

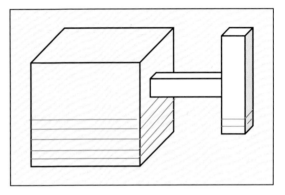

04.14 The object you want to print has to be cut into slices, the same height as the layers you want to print.

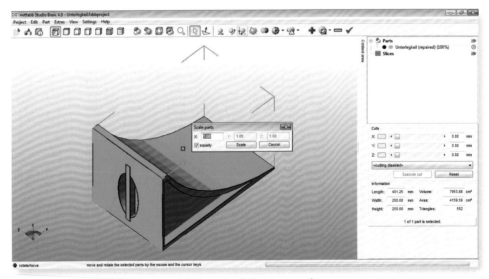

04.13 Netfabb can scale objects to the required size very easily.

into many slices, which are defined by the thickness of the layer that is to be printed with your printer. **(Illus. 04.14)**

I will show you the main functions of the slicing process using Slic3r, a freeware program that is also used by many commercial printer producers. In a later chapter, I will show some other slicing programs.

When you run Slic3r for the first time, it will ask for the main data on your printer. The program will use this information in the

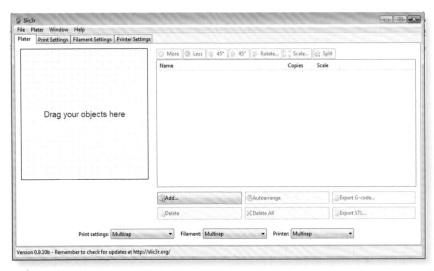

04.15 The main screen of Slic3r.

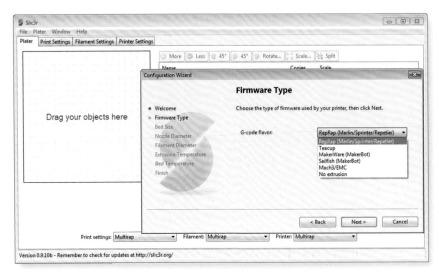

04.16 In the Wizard, you can choose the firmware your printer is using.

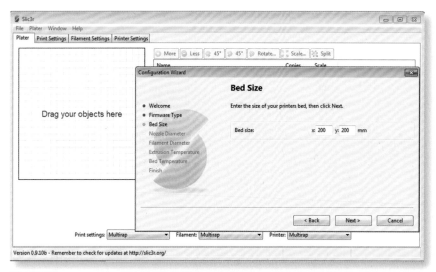

04.17 Then you can enter the size of the printing bed of your printer.

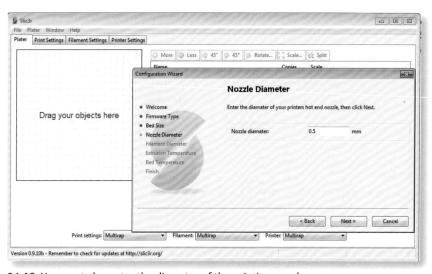

04.18 You must also enter the diameter of the printing nozzle.

future; you will not have to enter it every time you use the slicing process. If you change your printer, or there are any data changes needed, you can start this wizard manually and enter the new reference data.

Let's go through the wizard's questionnaire. The first question asks what firmware your printer is using. The firmware is the internal program which drives the printer. There are several types of firmware, and the slicing

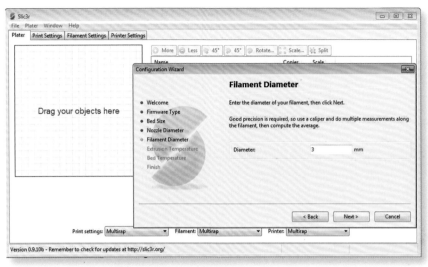

04.19 The filament diameter has to be typed in.

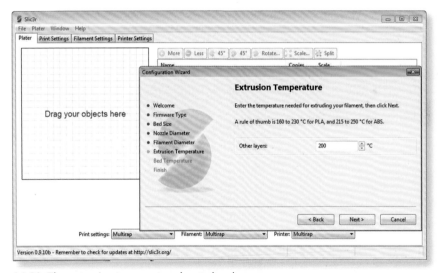

04.20 The extrusion temperature has to be chosen.

program has to know which one is installed to make sure the printer understands the g-code.

After you have entered the data, Slic3r asks for the size of the printing bed (e.g. the maximum printable size) and, after this, the diameter of the printing nozzle your printer has. This information is needed to ensure a trouble-free and high-quality printing result. **(Illus. 04.15), (Illus. 04.16), (Illus. 04.17), (Illus. 04.18).**

The following questions deal with the diameter of the filament being used and the temperature needed for printing. The printing temperature (which you can change later) is very important and must be found out by trial and error with the filament you are using. There are differences in the best temperature required for filaments from various manufacturers and also sometimes for the same filament from different batches. If there are any problems with the printing, changing the printing temperature will often solve them. For your first tests, a temperature of about 390°F for PLA will work; a good starting point for ABS is 430°F. **(Illus. 04.19)**, **(Illus. 04.20)**

The final question is about the temperature of the printing bed of your printer. Before I give you a usable temperature, I will say a few words here about heated print beds. One of the big problems in 3D printing is that the plastic has to stick to the printing bed. Even if this sounds a little bit odd, this is one of the key points for successful printing. Why this is so problematic is that we are printing with hot plastics that cool down very quickly. Due to this, the volume of the part shrinks and tends to lift off from the printing surface.

To prevent this, there are several methods in common use. Besides the use of glues or adhesive tapes, heated beds are the most common method. The beds are heated to a temperature that leaves the plastic warm so that it does not have a tendency to lift from the surface. After the printing process has finished and the bed has cooled down, the part can be removed easily with little physical effort and damage. This is one of the main advantages against the other techniques used to make the part adhere to the printing

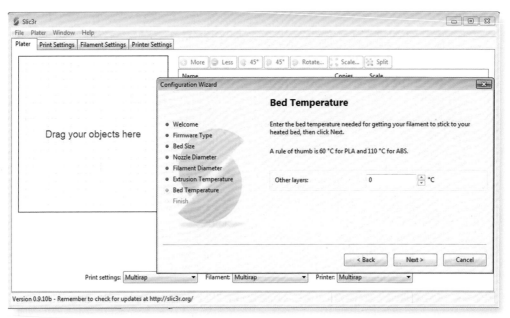

04.21 If your printer has a heated bed, you have to choose the correct temperature for this.

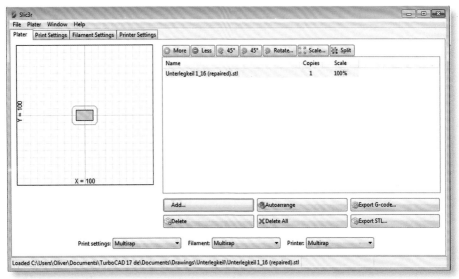

04.22 After you click on "Add," you can choose a file, and it will appear on the virtual printing bed at the left of the screen.

bed. So a heated printing bed is a very good method for ensuring a good printing result.

The normal temperatures used for the printing bed are 140°F for PLA and around 210°F for ABS (which does not adhere to the printing bed as well as PLA). (Illus. 04.21)

If you have worked through the wizard and saved the entries, you can prepare your first g-code. First load the STL file of the object you want to print. Click on "Add" at the lower part of the screen, and choose the file from your hard drive. (Illus. 04.22)

The part you want to print will be shown in the window in the upper left part of the screen in a view from above. You can make some changes here, like adding a second print of the same object that will be done at the same time, rotating the object, or scaling it up or down if you need a print in a different size. When working with Slic3r, you can arrange several different parts from different files so

that they can all be printed at the same time. (Illus. 04.23)

Go to the lower part of the screen to select the printing settings. Select the filament to use and the printer you want to use for your print. The filament and the printer will be normally be the same for every print, but the printer settings can change from print to print. This is because on this screen some very important data for the quality of the print is defined. You can define and save profiles for the different kinds and qualities of prints you wish to do. There are a lot of parameters that can be changed for defining the kind of print you want. I will go into details for some of them. As you work with your printer, you will find some more ways to get even better results, but for most prints it will be enough to adjust the points described next. A very nice feature of Slic3r is that if you hover the cursor at one point, it will give a short description of

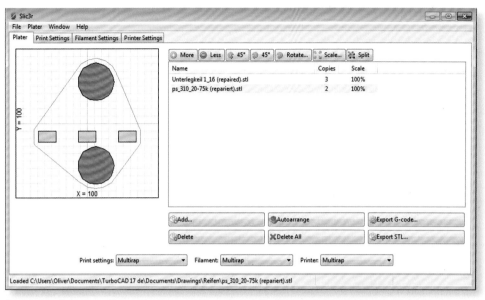

04.23 By clicking on "More," you can add copies of the first part; by clicking on "Add," you can put other parts on the same plate for printing at the same time as the first part.

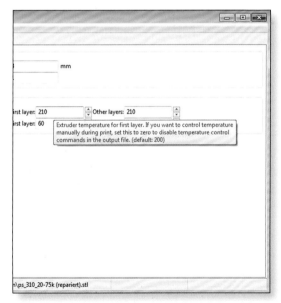

04.24 A nice feature of Slic3r is the help windows that pop up if you put your cursor over a box.

the parameter and the effects that changing it may have on the print, and it also gives hints for the best value for several different circumstances. **(Illus. 04.24)**

If you go to "Print settings," you will find some sub items that you can work through. The first selection, "Layers and Perimeters," is very important for the printing quality and especially for the finished surface of your print. As an FDM printer puts the material on in thin layers, one over another, there will always be some visible steps on most parts of the surface. You can minimize them by making some adjustments, but not remove them totally. Of course, the thinner the layers are, the less visible the steps will be, but this will only work if your printer is adjusted very carefully to do such a fine print. Normally printers will work well in a range from 0.1 to

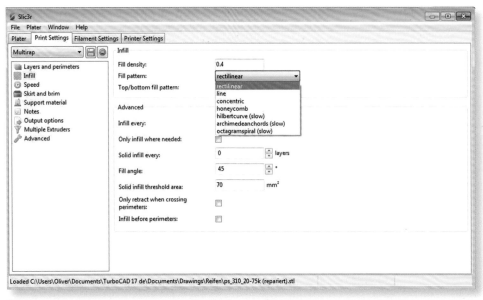

04.25 In this dialogue, you can enter the thickness of the layers you want to print.

0.5mm. Below this resolution (which some manufacturers are advertising), it will not be so easy to print. You can try printing such thin layers when you have some experience. You can change the layer height at the top of this table. **(Illus. 04.25)**

The first printing layer is normally printed a little bit thicker than the following layers because this lets the part adhere to the printing bed better and gives more stability. You can set the value for this, but normally when you print thin layers, the first layer should have a thickness of about 150%.

The next selection you define will be the "Infill" of your print. Your printer can print complete structures, hollow shells, and everything in between. Printing hollow shells will often not work because they need a supporting structure inside. Also, large prints will take a lot of time and material, so the best way to proceed is to print a partially filled body. How much the print should be filled in will be defined by a number between 0 (hollow) and 1 (massive filled). Normally a value for the "fill density" of 0.3-0.5 is best. **(Illus. 04.26)**

The next selection, "Speed," is very sensible, and, if you are not experienced, you should use the default values. When you have more experience it may be helpful to check out some other speeds; high speed may result in poor results as will a slow speed. Lower the printing speed if you realize that your printer is not stable enough to print at high speed and is vibrating, thus making the printing result poor. **(Illus. 04.27)**

"Skirt and Brim" is the next option you have to deal with. This means that before the printing of the part starts, the printer will put a line of extruded material around the base of the printed part. This ensures that the extruder nozzle is completely filled with

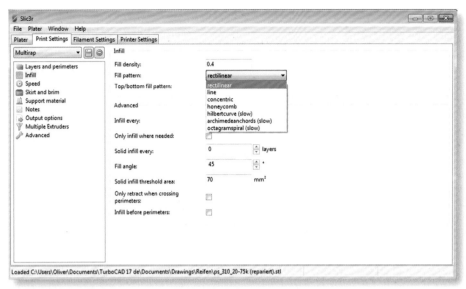

04.26 Under "Infill," you can choose the filling density and the pattern with which the object should be filled.

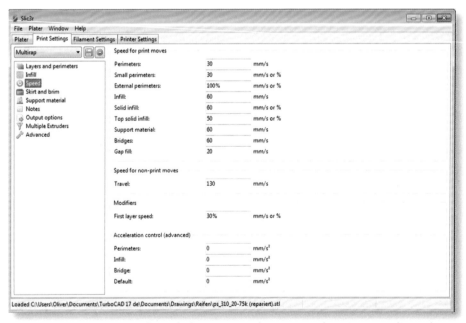

04.27 You can experiment a lot with the print speed to improve the printing quality and time, but it is a very sensitive option.

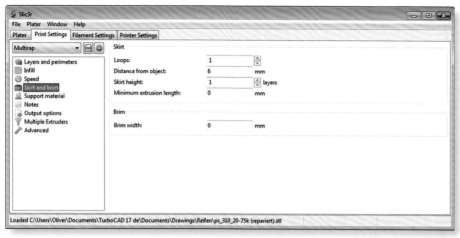

04.28 There are not many options for the skirt and the brim.

material before beginning the real print. This should be enabled, because it does not use much material and makes the initial print layer better. (**Illus. 04.28**)

The next option is a very important one called "Support material". Because a 3D printer is printing material in layers, one on top of the other, it needs a base for every

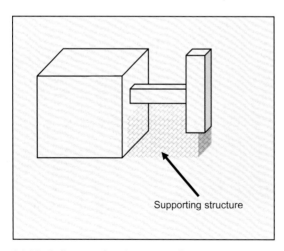

Supporting structure

04.29 A diagram showing a typical support structure.

layer to print on, or the material will be unsupported. (**Illus. 04.29**)

Normally the nozzle will lay the material on the layers previously printed on the printing bed. Many printed parts will have an overhanging section that has to be printed in midair, which is, of course, impossible, so we need a support, e.g. printed stanchions, on which the printer can lay the material for the overhanging parts. These support structures must not be as massive as the real printed part so that the structure can be removed easily from the part when it is no longer needed.

Fortunately, we do not have to construct the supports on our own, because the slicing programs will construct them where they are needed. Normally you will not have to change anything in the default values, because they work very well. You only have to decide whether you want to print the supports or not, because some constructions will not need any support at all. (**Illus. 04.30**), (**Illus. 04.31**)

Another decision you need to make is if you want to print a so-called "raft". The

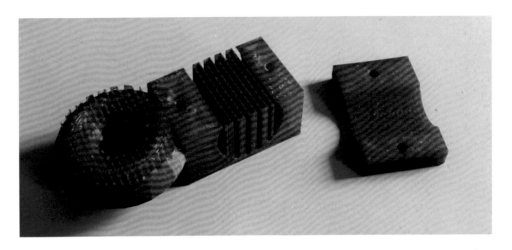

raft is an extra layer, like a support, that is printed before the normal printing starts. The reason for this is that some constructions do not have a first layer as a base that is big enough to adhere to the printing bed. For such delicate parts, the raft makes the part adhere to the printing bed successfully. **(Illus. 04.32), (Illus. 04.33)**

The remaining settings, "print settings," do not need altering for your first prints, but you might find them useful when you are more experienced.

Some more adjustment is possible for the "filament settings." Although we have already entered some of this data into the wizard, it may be helpful to adjust some data here. The first selection is the "diameter" of the filament. Even if the normal filament diameters are 1.75 mm and 3mm, some producers vary these dimensions a little. Often 3 mm filament is only 2.85mm in diameter. You should check the diameter of the filament with a caliper, because if your slicing program is calculating the print using the wrong diameter of filament, the print will have a poor quality because the printer will use too much or too little material.

Another selection is the "extrusion multiplier," which may be useful. Not every filament has the same density of material when extruded, so a print with a filament with low material density will not have the same dimensions as a print with a filament of high density. Adjusting this selection can compensate for this effect.

Even more important is the adjustment of the temperature, which is easy to do here. There will be a difference in the temperature

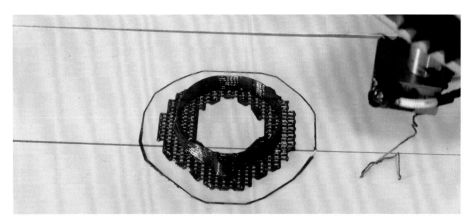

04.32 A raft increases the adhesion of a printed part to the bed; this is especially important for small and fragile parts.

between the first layer and the following layers, because you would normally print the first layer with a higher temperature to ensure good adhesion to the printing bed, making a good surface for the following layers.

Sometimes it is necessary to change the temperature because different filaments (and even supplies from the same supplier) may need a different temperature; vary this point if you are not satisfied with the result of your printing. Even the ambient temperature may influence the result of the printing; if it is too low or too high, you may have to compensate it with an adjustment of this selection.

You could also, if your printer allows this, vary the temperature of the printing bed. Even this can be varied between the first and the following layers. Sometimes it may be

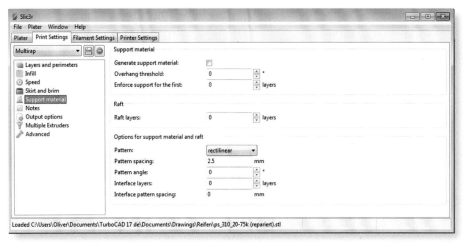

04.33 The parameters for the support and raft can be entered on this screen.

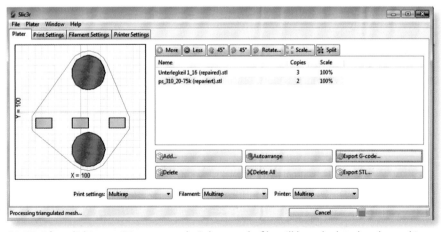

04.34 There are only a few parameters entered in the filament settings, but these are very important.

useful to let the printing bed have a higher temperature for the first few layers to let the print stick better. This is also a point that can be influenced by several parameters, e.g. the filament used and the ambient temperature, so you should use whichever value works best for you. **(Illus. 04.34)**

The last step will be to click the button "Export g-code." After you select where on your hard drive it should save the code, the program will calculate the g-code and save it. That's it, all done! Now you can use the resulting g-code to drive your printer. **(Illus. 04.35)**

04.35 After clicking on "Export g-code," the g-code-file will be calculated and saved in the entered folder.

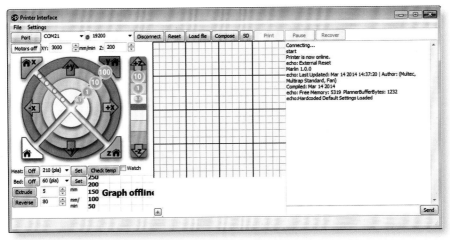

04.36 The screen of a simple printing program called Printrun.

Now load the g-code to your printer. The printer has to be driven, and it needs a program for this. Normally a printer will be delivered with a driver program, where you have to load the g-code from your slicing program. **(Illus. 04.36)**

Often the printers are delivered with a package of programs that not only contains the driving program but also a slicer that

works in the background. I will give you examples of this later in this book.

At this point, I would like to give a short introduction to a puristic driving program, called "Printrun," which is a good example of the general method of operation of driving programs.

After you start the printing program, you have to connect it to the printer, so it needs

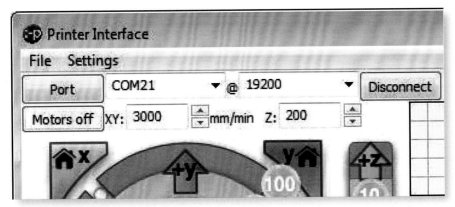

04.37 It is important to enter the correct port and the correct baud rate to make the connection between the printer and computer successful.

some information about the computer port you are using. This is essential information, and you will get it when you first connect your printer with your computer. Sometimes finding the right port and connection parameters is a little bit tricky, as they may be in conflict with other (normal) printers or equipment. The manufacturer or distributor of your 3D printer should give you good and helpful support and so troubleshooting should be no trouble at all. I have installed numerous 3D printers on a couple of computers that worked straight away or after a short bit of troubleshooting. I am not counting the ones where the non-working connection was my mistake…

In Printrun, you have to choose the port and the baud rate at which the communication between the computer and the printer is done. Normally you can scan the ports on your computer by clicking on the button "Port" at the left-hand side. The new possible ports will be shown in the pulldown menu of the program. By clicking on "Connect," the computer and printer will be connected if everything works. **(Illus. 04.37)**

After this connection, you can use the printer. In every program, you can drive each of the three axes by hand and move the printing table and/or the printing head to every position you need, such as for cleaning, maintenance, or repairing. For your first print, you should bring the printer to its home position (normally indicated by a house symbol) and then prepare the first print. **(Illus. 04.38)**

The printing head and, if your printer has one, the heated bed need a little time to reach the right temperature, just like your oven in the kitchen if you want to bake something.

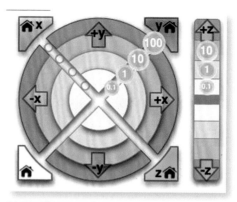

04.38 This console allows you to move the stepper motors manually and bring the axes to their home position by clicking on the home symbols.

Even if the printing program checks if the right temperature has been reached, it is a good idea to preheat them on your own.

In Printrun, you have the option of adjusting the temperature for printing in the lower left corner of the console. For the printing head and the bed, you will have three temperatures to choose between: 32°F for off, a temperature for ABS, and one for PLA. Sometimes you will have to fine tune these temperatures a little bit, because there may be variations between the filament from different manufacturers and even between batches of material from the same supplier.

Normally a temperature for the print head of 475°F for ABS and 410°F for PLA, with a bed temperature of 210°F for ABS and 140°F for PLA, will work well.

To prepare your print, click on "set" after you choose the temperature of the material you are printing. Then check the temperatures: watch the rising temperatures in the graph right beside the buttons after clicking on "Check temp" and tick the box

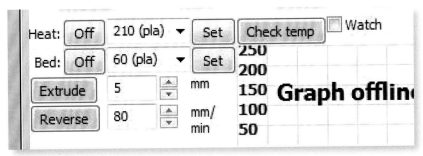

04.39 In this area of the console, you can enter important information about heating and so on.

"watch." Now you will see a graph of the required temperature of the bed and printing head and the current temperature. When the index values and the actual values match, the printing can start. (**Illus. 04.39**)

As reaching the required temperature normally takes a little time, load the g-code of the object you want to print while you are waiting. After clicking on the button "Load file," you can search for the g-code you produced for your printer on your hard drive.

If you use different printers, be sure that you load a g-code file that is made for the chosen printer, otherwise it will not work, and you will get an error message. (**Illus. 04.40**)

After the file has loaded, you will see a graphic of the printed part lying on the virtual printing bed. On the right side of the screen, you will get some information about the print. Some slicing programs (like Slic3r) calculate the required length of filament, but others may not. It is not really important

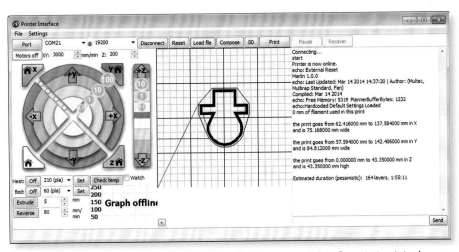

04.40 After you have loaded a file, a picture of it appears and data for printing it is shown at the right-hand side.

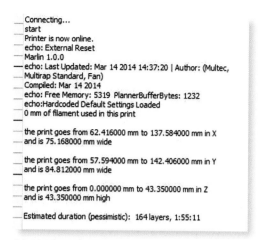

```
Connecting...
start
Printer is now online.
echo: External Reset
Marlin 1.0.0
echo: Last Updated: Mar 14 2014 14:37:20 | Author: (Multec,
Multirap Standard, Fan)
Compiled: Mar 14 2014
echo: Free Memory: 5319  PlannerBufferBytes: 1232
echo:Hardcoded Default Settings Loaded
0 mm of filament used in this print

the print goes from 62.416000 mm to 137.584000 mm in X
and is 75.168000 mm wide

the print goes from 57.594000 mm to 142.406000 mm in Y
and is 84.812000 mm wide

the print goes from 0.000000 mm to 43.350000 mm in Z
and is 43.350000 mm high

Estimated duration (pessimistic): 164 layers, 1:55:11
```

04.41 The value of the z-axis must start at 0.000000 mm. The time shown will often be extremely accurate.

information, but may be useful if you only have a little remaining filament and you are not sure if it will be enough for your print. The next information is about the dimensions of the print area in the three axes. Even if it is a little bit confusing with all the decimal places, this information is very useful. You can see here, as a final check, if the part fits onto your printing bed and especially if the z-axis movement is sufficient. **(Illus. 04.41)**

To get a correct print, the printing should start in the Z-axis at a value of 0.00 mm when it will start correctly on the printing bed. Sometimes, during the construction of parts, it will not be oriented so. If the Z value is negative, the printer will print the first layers until it reaches the neutral point in one layer, which means the printing will not work.

If the print is starting at a positive height above the printing bed (except for some special uses), it will print into midair, which means the extruded material will fall down, also not a good basis for a usable print.

The last information given is about the estimated time that the print will take, which is very interesting. Surprisingly, the calculated time often correlates with the actual time needed for printing very precisely within an acceptable deviation. Also indicated are the number of layers that will be printed until the part is finished.

When the temperatures of the printing nozzle and bed are reached, you can load the filament. How to do this will be described in your printer manual.

After the filament is loaded into the printing head, you should fill the printing nozzle with material. Click on the button "Extrude" at the left-hand side and the filament will be fed into the nozzle and melted. After enough material is molten (this may take several clicks on the button) a thin thread of melted plastic will come out of the nozzle. Careful, this material will be very hot after it

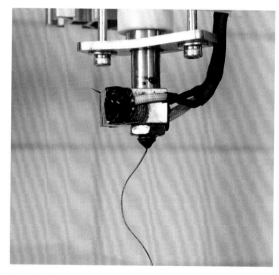

04.42 When the printing nozzle is hot enough, a thread of molten plastic can be extruded.

4.43 Watching a 3D printer working is always fascinating.

has left the nozzle! However, it will cool down surprisingly quickly, allowing you to remove it from the printing bed. **(Illus. 04.42)**

Now everything is ready for your first print, but one moment! One last check should be done; sometimes the movements of your printer could be blocked by objects that should not be in the way. Tools, residues of filament, or even a coffee mug can interfere with the printing process or cause damage to your printer. So take a last look at your printer to check if all the axis movements are free.

If so, click on "Print" and (sometimes with a short delay due to some temperature fine tuning) the print will start as if by magic. Layer after layer will be printed, and the object you await will be growing. This will work automatically and you will (at least for the first few prints) be fascinated watching at the process. It is amazing! **(Illus. 04.43)**

05.01 Printing a two-color traffic cone from www.thingiverse.com
(thing:21773, user CocoNut).

5

Data for printing: Thingiverse, etc.

Buying or building a printer and being able to use it is one thing, but you must have the three-dimensional data file of an object to be able to print it. This is, for most people, the biggest challenge. Three-dimensional construction is not black magic if you take a little time to learn it, but it is not as simple as making a normal two-dimensional technical drawing. In the next chapter, I will give you an example for a simple 3D construction, but first I will give you a brief overview about how you can get usable data for several objects you can print without spending several hours in front of a computer. **(Illus. 05.01)**

3D printing would not have made such enormous progress without the internet. A big community now exchanges their experience over borders and continents and has created the basics, not only for self-build printers

but for nearly all the commercial products in this area.

Not only is the hardware and software the result of this great number of enthusiasts, an intense community is working on 3D construction and is making it possible to download (mostly free of charge) an enormous number of files for such constructions.

There are several platforms for this exchange (and there are more being added), but the most frequented is www.thingiverse.com, a website that is provided by Makerbot, probably one of the companies that has had the biggest influence in making three-dimensional printing for the home user possible.

On Thingiverse, anyone can join the community and upload or download 3D files. Of course, they must be compatible with the rules of the website. This is a big advantage

05.02 This nice soap dish was made from a file on Thingiverse. *(thing:135154, user brackett27).*

of the website and also a problem. Even if the people behind the website are checking the uploaded files from time to time, it is no guarantee that the files are printable. The variety is great; from fascinating and extremely accurate constructions that are printable without any work from you, to vast data constructions that will not print even with a lot of subsequent treatment. Almost anything is possible. **(Illus. 05.02)**

Using Thingiverse is quite simple. After you choose the object you want to print, you will download the files for it, and these are usually available as STL data. Sometimes there are some special data formats, which are not as easy to use, so be sure that you can open the files if they are not in STL format. Of course, you should check if the files are clean, so check them if this is not done automatically with your antivirus software.

After this you should check the STL files with netfabb for correctness and for the right orientation of the object. As you saw in the previous chapter, there could be a negative value for the Z-axis. If there are any problems, correct them. Now you can slice the part with your favorite slicing program to make the data usable for your printer and then print it. **(Illus. 05.03), (Illus. 05.04)**

If you search for a part for your printer, a spare part for something in your home, or even a nice decorative object, you will find a lot to print on Thingiverse and other websites. Please remember, be fair! The files you find on any of these exchange communities are the intellectual property of the person that made them available. So make it clear to everybody that your print is based on a file from somebody else and that it was not your idea or design. This also means that you

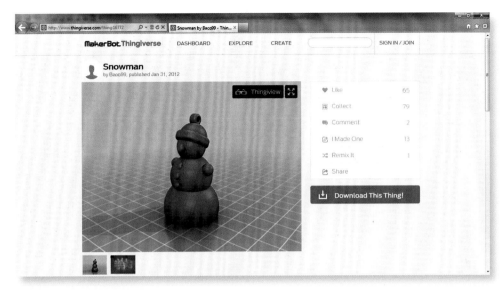

Above: 05.03 A nice little snowman as shown on Thingiverse...

Left: 05.04 ...and the file printed, one in a more unusual black color. *(thing:16772, user Baco99)*

should not sell prints of parts that are based on the intellectual property of someone else.

Of course, Thingiverse is not the only exchange platform that exists, but it is probably the biggest and the one with a file for nearly everything you could wish to print. Other websites do not have the same number of files or are only dealing with a limited kind of very special data. I will introduce you to

some of these that I find interesting; if you search for more, you will find a lot of them!

One of these websites is www.youmagine. com, a site that is powered by Ultimaker, a Dutch manufacturer of the 3D printer of the same name. On this platform you will find, in a very similar way to Thingiverse, a lot of interesting designs; some of them you will also find on Thingiverse. The number of files is not as big as on Thingiverse, but it is increasing steadily. The usage of these websites is simple and self-explanatory. (**Illus. 05.05**)

If you like technical printouts (or need them for your business), the website www.traceparts.com is very interesting. In fact, this website is designed for engineers, and you can download 3D CAD data of parts from several companies. Because of their normal use, not all of the files on this website

05.05 The home screen of www.youmagine.com

are actually useable for 3D printing, so do check them before you try to print them. **(Illus. 05.06)**

Another really interesting website for 3D files (which were not originally designed for 3D printing) is grabcad.com. There are thousands of mostly technical files for cars, planes, etc. With a little checking and work, you could use them for 3D printing. **(Illus 05.07)**

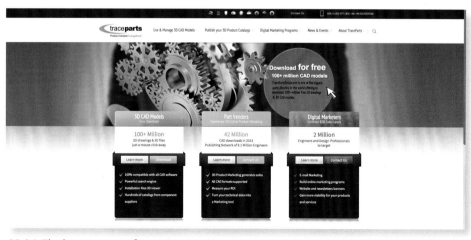

05.06 The home screen of www.traceparts.com

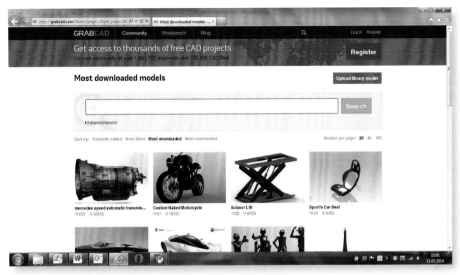

05.07 The home screen of grabcad.com

Finally, I want to give three more examples of very special downloads. The spaceships, satellites, rockets, and space-capsules that were used by NASA are fascination objects of technical history. Now NASA makes it possible to print many of them on your home 3D printer as they are providing a lot of 3D data on their website, found at http://nasa3d.arc.nasa.gov.

05.08 Even NASA is providing digital 3D data of lots of their equipment (http://nasa3d.arc.nasa.gov).

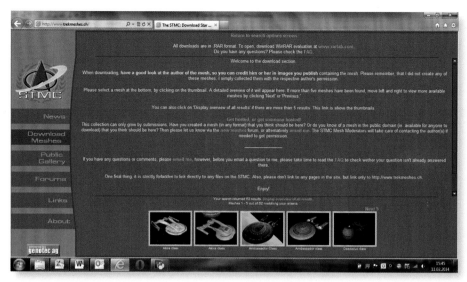

05.09 The perfect page for Trekkies is www.trekmeshes.ch

Because they are provided as 3ds-files, you have to convert them into STL-data, which is possible with netfabb. (**Illus. 05.08**)

Also in space, but in the far future, are the 3D files you can download from www.trekmeshes.ch. Here you will find nearly all of the spaceships and other stuff from Star Trek! It is really astonishing the way they are modeled; you could bring a Star Trek universe to life. (**Illus. 05.09**)

05.10 Honda also provides digital data of some of their concept cars (www.honda-3d.com).

05.11 Archive 3D is also a good platform, but not all of the files are printable at once (www.archive3d.net).

But let us get back down to earth. Honda, the Japanese industrial giant, has opened their archives of concept cars, and on the homepage at www.honda-3d.com you can download some very good 3D models of some interesting and uncommon cars that will never be built. It is interesting that such a big company not only makes their 3D data accessible at all, but that they pre-process it for 3D printing at home. (Illus. 05.10)

There are some more platforms for three-dimensional files which could also be used for 3D printing. But these are mostly specialized, because most of the files are not drawn for 3D printing and often they have to be worked over or are not printable at all. Some platforms include some free-of-charge models, but others have to be paid for. Also, a couple of new platforms have been launched that are selling 3D data for printing. So take a walk through the internet for finding new and interesting 3D files.

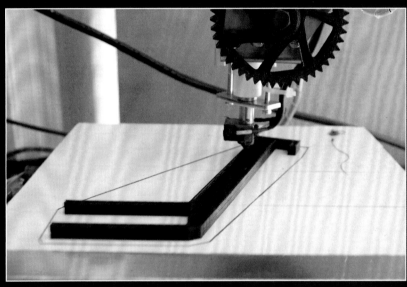

06.08 After slicing, the tablet holder is ready to be printed. *(see page 72)*

6

Your own construction

As we saw in the previous chapter, there are lots of files for parts that you can print without any construction work from you. So why should you make the effort to make your own construction? Because this is the real advantage that 3D printing at home brings: you can produce exactly the parts and products that you, and maybe only you, need. You can construct a part exactly right for the purpose you need it for. You will have the exact part to solve your problem.

Let's get a little bit into 3D construction. There are several possibilities for constructing something in 3D. If you want to construct something technical to exact dimensions, you will need another program, CAD software (Computer Aided Design). If you want to work more artistically and make some decorative objects, artistic software like Blender (www.blender.org) will be better.

3D construction needs a lot of work. First you must learn how to use the software and get a feeling for constructing in a three-dimensional space. This thinking is one of the biggest challenges; if you can't think three-dimensionally, you probably can't construct three-dimensionally. Take your time learning how to use your construction software and try out all the functions your software has. 3D software is always very complex, and it will take quite a lot of time and constructions to learn even the basic functions.

This short chapter does not give you a complete introduction to 3D or CAD construction; there are many books available to teach you the basics of CAD. I will just give you a short guide to basic 3D design.

Shortly after your family discovers that you own a 3D printer and what is possible with such a machine, you will get the first requests asking if it would be possible to print this or that. Do not be displeased about this, but use these requests for learning new construction skills.

One such request was for a holder for a tablet computer for our kitchen. Print cookbooks are outdated; today, you search for a recipe on the internet. However, it is not advisable to lay your tablet on the kitchen worktop, especially not with water, oil, and other stuff beside it.

So the task was to construct a removable holder for the tablet that could be attached to a kitchen shelf. The description of this construction process is easiest to show in the photos that follow. The kitchen shelf has a thickness of 20mm, and the tablet is 205mm × 160mm and 10mm thick. The printer I used for this part has a maximum printable height of 150mm, and, as I did not want to print with too much supporting structure (which is a lot of work to remove), I decided to print the tablet holder on its side. This is one thing that you have to consider before you construct a part that you want to print; think over how you want to print it and the best way to print it. This will make things much easier.

For this tablet holder, I choose a very simple form of construction, which is possible in all CAD programs. You draw the outline of the object you want to produce and extrude this to the correct height. You will get a large part that you can transfer to a printable STL file. **(Illus. 06.01), (Illus. 06.02), (Illus. 06.03), (Illus. 06.04), (Illus. 06.05), (Illus. 06.06), (Illus. 06.07), (Illus. 06.08), (Illus. 06.09)**

This was a very easy construction job that only required extruding a 2D design into 3D.

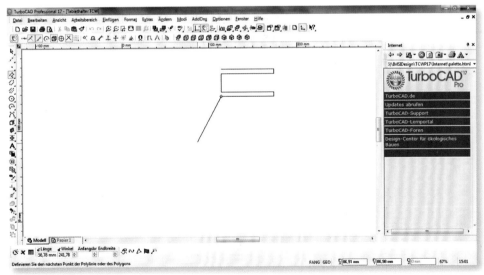

06.01 In this kind of 3D construction, you draw the contour of the object you want to construct (and print).

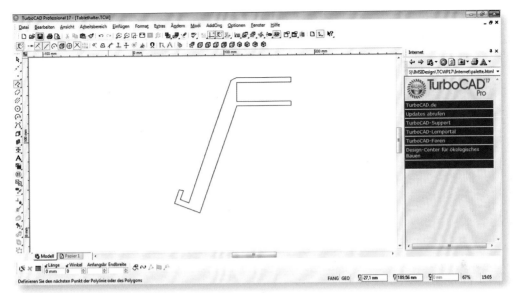

06.02 This is the complete contour of the tablet holder.

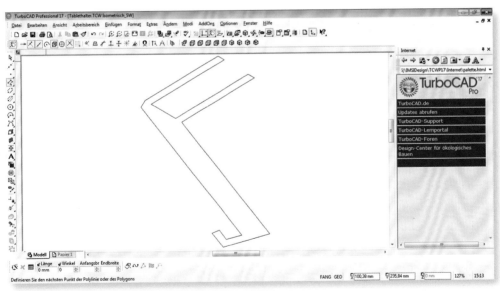

06.03 After this, we change to a three-dimensional view.

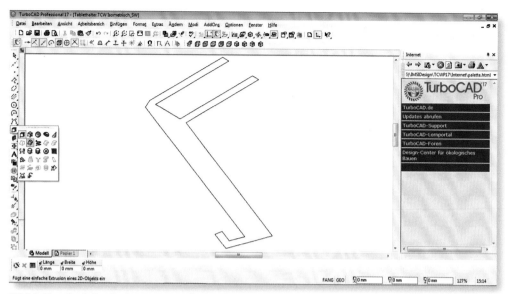

06.04 Now look for the icon to extrude the drawing.

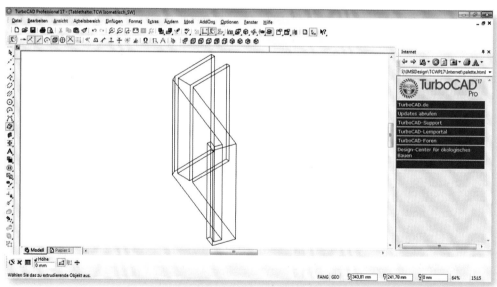

06.05 The contour has now been extruded.

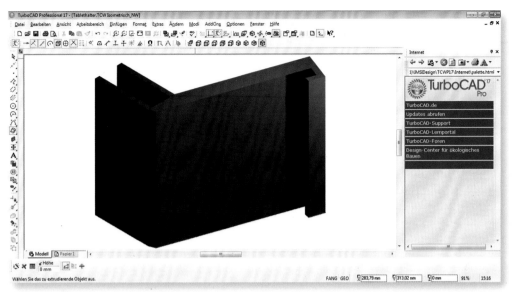

06.06 Now you can render the object in color, and you have a nice impression of how the finished object will look.

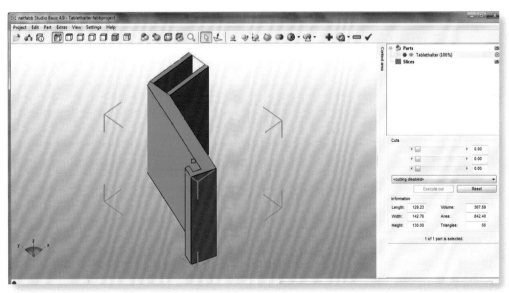

06.07 In netfabb, you can control and repair any problems.

06.08 After slicing, the tablet holder is ready to be printed.

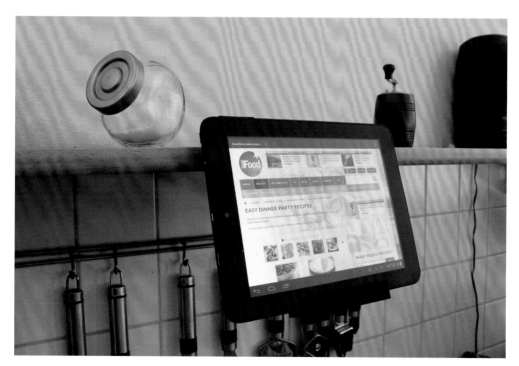

06.09 With this tablet holder, you can use your tablet as an interactive cookbook.

The use of 3D parts to build your construction is a little bit more sophisticated; it is a bit like playing with virtual toy blocks.

The main principle is that in such a construction you have to think three-dimensionally. With a little bit of practice, this will not be a problem. In every 3D CAD program the principle of this is the same. An object like a cube, cone, or sphere is melded with each other or its volume is eliminated from the volume of another object.

For clarification of this abstract procedure, let me show you some more photos of a construction. **(Illus. 06.10), (Illus. 06.11), (Illus. 06.12), (Illus. 06.13), (Illus. 06.14), (Illus. 06.15), (Illus. 06.16), (Illus. 06.17), (Illus. 06.18)**

Even if this is hard to describe, once you have understood the basics and tried it out, it will not be as complicated as it sounds.

CAD is not the only method for 3D construction, but it is my favorite, especially if you want to construct usable technical parts. If you want some decorative objects or another kind of artistic object, other software will be better to use.

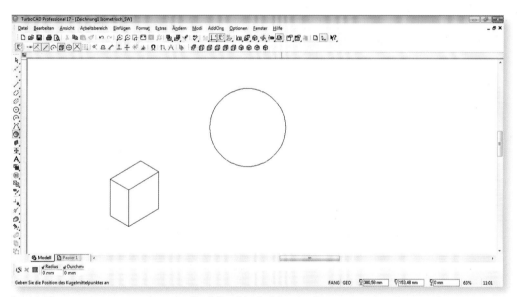

06.10 First you have to produce the "building blocks" for the 3D construction; here we have a cuboid and a sphere.

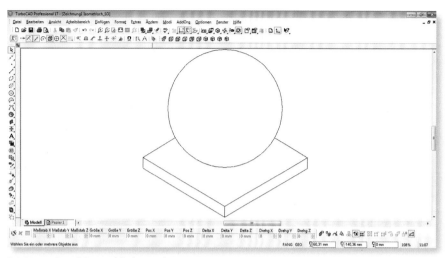

06.11 Both will be melded together.

06.12 Using the icon highlighted in this picture, both parts will be melded together.

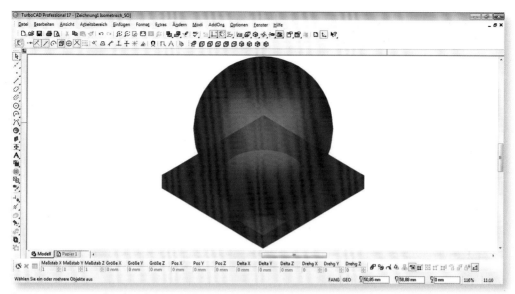

06.13 You can render the object to get an impression of the printed object.

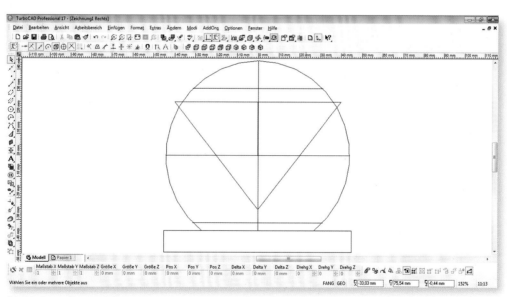

06.14 As well as melding parts, you can remove them from each other. Here a cone is removed from a sphere.

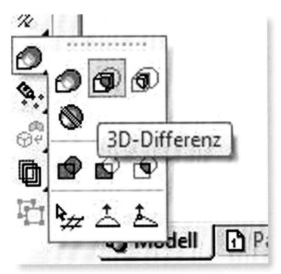

06.15 With the highlighted command shown here, you can remove one object from another.

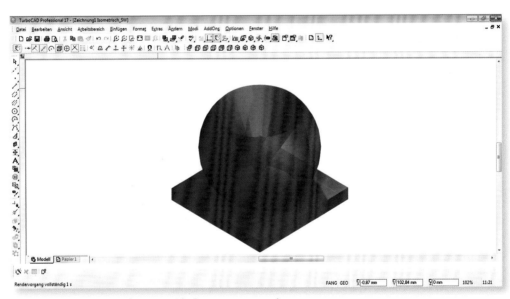

06.16 Here we can see the part with the cone removed.

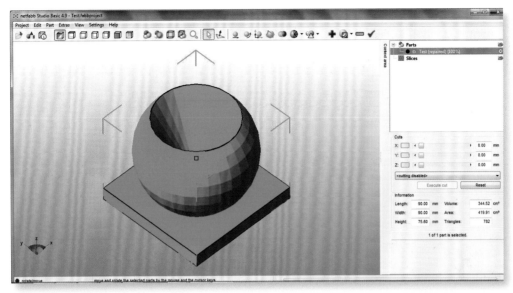

06.17 The part is checked in netfabb...

06.18 ...and this is the printed result.

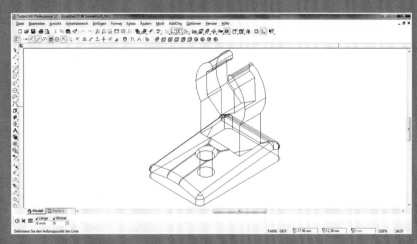

07.01 For technical drawings like this, CAD software like TurboCAD is perfect.

7

Software for construction and preprint

The market for CAD software is nearly unimaginable. Freeware programs up to programs costing tens of thousands of dollars are available. For our use, simple CAD software is fine, and it need not even be the latest edition. I am using an older version of TurboCAD, which I find very easy to use (after some time spent learning it); it is complete enough for my requirements. These older versions only cost a few dollars, although the latest version is around several hundred dollars to purchase.

Make sure you buy a CAD package that suits your needs. One absolutely essential thing is that you can construct your part in 3D; not all CAD programs can do this, so make sure that three-dimensional drawing is possible with the CAD package you have chosen.

It is also very important that you can save your construction in a format that is readable in other programs. Some CAD programs, especially cheaper ones, may only save the drawings in an unusual format so you can only open it with the same program you drew it in. This is not what you want. Be sure that you can save your construction in several formats like, for example, *.dwg, *.3ds, or *.obj, which can be imported by netfabb or other software that you need to prepare the files for the 3D printing process. Ideally the program will save the file as a readable STL file; many of the better CAD programs have this option. (**Illus. 07.01**)

A very nice free CAD program is FreeCAD (www.freecadweb.org), which allows a lot of construction and has good intuitive handling. It is an open source project and allows a lot of customization by the user. (**Illus. 07.02**)

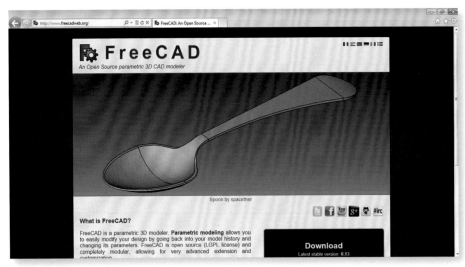

07.02 A very nice free tool for CAD is FreeCAD.

Not all users of 3D printers will want to print linear technical objects, so it is important that they have the right tools. A lot of parts are sculptures, and for them, CAD is not the right tool. There are a lot of free programs for making your own sculptures, etc.

Most of them are normally used for digital art, digital animated films, or computer games, but they are very useful for us as well.

Meshmixer a very powerful free application that has gone through many upgrades since being acquired by Autodesk. The software

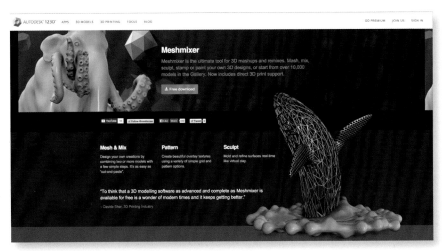

07.03 Meshmixer is a powerful, helpful free application.

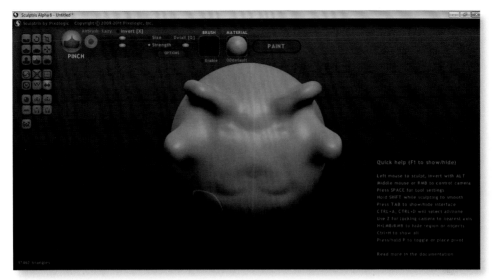

07.04 With Sculptris, you can work a design on your computer like you can work clay with your hands.

offers not only the ability to sculpt and colorize, but it also can do many of the diagnostic and repair processes found in netfabb. As part of a joint agreement with Makerbot, the software has a direct output to their Makerware software, allowing for direct design to print in one software package. (**Illus. 07.03**)

Sculptris (from www.pixologic.com/ sculptris) is quite easy to use. It is like you have a virtual clod of clay that you can form with your computer mouse, just like you can form clay with your hands. I find this program really fascinating, even if I am not a real specialist, and it is a very nice tool for anyone who wants to work artistically. You can export your files into the OBJ format, which is easy to use with netfabb. (**Illus. 07.04**)

A program that is not really CAD-software but is also not a real sculpting software is Sketchup. Formerly known as Google-Sketchup, now bought by Trimble and published under the name Sketchup 2013 (www.sketchup.com), it is an easy-to-use design software that has a lot of uses. For non-commercial use, the version "Sketchup Make" is free of charge and has all the features that you will need for your hobby constructions. (**Illus. 07.05**)

Autodesk, the provider of the professional and expensive CAD software AutoCAD, offers a variety of more or less easy-to-use apps (www.123dapp.com) that are usable for constructing files for 3D printing. 123D Sculpt and 123D Design are easy to use to make simple constructions or designs for your 3D printer. (**Illus. 07.06**)

If you have your construction, you should check it before you slice it and try to print it. I have already written about my favorite program for doing so, netfabb (www.netfabb.com). The version called netfabb Basic is free of charge and has all the

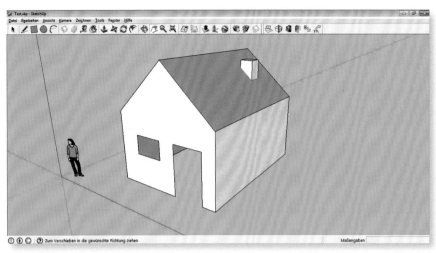

07.05 Sketchup is nice and easy to use for constructions.

features you need. You should check your files for integrity, orient the part in the right direction for printing, and at the right level, you can scale it to the right size and convert it, if needed, to the right format. Personally, for my 3D printing, netfabb is one of the most important tools. **(Illus. 07.07)**

If you have a very unstructured STL file, e.g. from a 3D scan, it may be necessary to check it using special software. One software that can do this is Meshlab (meshlab.sourceforge.net), a special open source project that is very helpful for such cases. **(Illus. 07.08)**

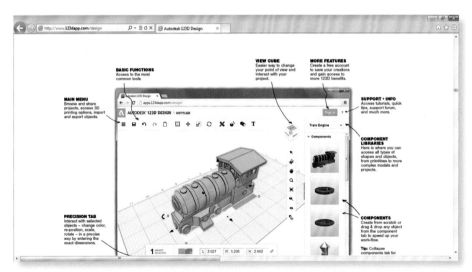

07.06 123D Design by Autocad (www.123dapp.com) is a nice and free construction tool.

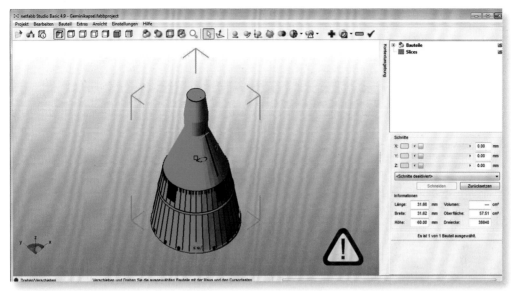

07.07 As we have seen already, netfabb is quite a powerful tool.

07.08 Meshlab can be very helpful, too.

08.01 The slicing software Skeinforge is very
powerful, but not so easy to use.

Slicing software

As we have already seen, the object you want to print has to be prepared for printing. The printers are printing the objects in slices, so the constructions have to be cut into these slices by the software, just as you cut bread for breakfast.

However, the software needs some information to do this. First it needs to know what kind of printer (e.g. what kind of driver) you have and how you want to print your object.

There are several slicing programs on the market. Normally they are freeware, programmed by enthusiasts that are doing a great job making superb software free of charge; many thanks to them for that!

I would like to introduce three programs that are popular with many users. Most printers offer the ability to use different slicing programs, but some are implemented using a dedicated slicing program in their printing software that you must use. Some printers have a special driving screen, which you do not see directly, that hides the slicing software being used. Often you will find in the background one of the slicing programs mentioned in this book.

One of the most popular slicing programs, called Slic3r, has already been mentioned previously in Chapter 4, "The first print." Slic3r is a very easy-to-use program that has a lot of options, but it does not always give the best result possible because the options to fine tune it are sometimes a little bit limited.

I would now like to introduce a program that is not as good-looking on the screen as

Slic3r is, but which has many options to fine tune the results. It is called Skeinforge, and you can download it at http://fabmetheus.crsndoo.com.

Because it has a wide variety of tuning options and a matter-of-fact screen-layout, it may be a little bit confusing to learn, but if you work a little bit with this program, you will soon find it easy to use. **(Illus. 08.01)**

Skeinforge is not only usable for making the g-code for 3D printers, but it can also be used for milling machines and other machine tools. For 3D printing, only the "Extrusion" option is interesting. Like most other slicing programs, in Skeinforge you can create profiles for different printers and printing processes. For example, you could create profiles for printing with or without support, for thin or thick layers, and for other specialities. You can save these profiles under specific names and call them up when you need them. Some manufacturers

of 3D printers deliver ready-to-use profiles for Skeinforge (as for some other slicing programs); you can usually download them from their homepages, or they may be included with the software for your printer. These profiles are often very good and are a good base for fine tuning your printing results. You can use a basically good working profile and modify the options in small steps. Make sure you have a backup of the original profile to go back to if you tuning does not work!

Skeinforge has very good online help that you can activate by clicking on the question mark next to most of the option buttons. I will now guide you through the most important features of Skeinforge.

Carve

Under "Carve," two options are very important. The first is "Layer Height," which is the thickness of the layer that the printer will put on the object. It is indicated in millimeters, and although it is normally a thin layer that lets us print finer details, it is not always necessary to print with the thinnest layer, especially as the printing time will increase by a multiple of the extra layers printed.

The second option here is "Edge Width over Height." This is important because the printer is not laying a round thread on the object, but rather a thread that is wider than it is high. For example, if this value is 1.8 and the layer height is 0.4 mm, the thread will have a width of 0.72 mm. This value is important for the fineness of the surface and for getting a stable outer skin. You should experiment with this value when you have some experience. **(Illus. 08.02)**

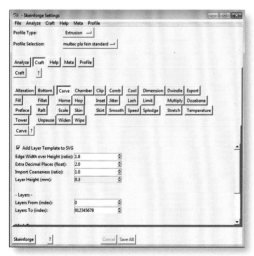

08.02 "Carve" is a very important option when working with Skeinforge.

Chamber

The next option is "Chamber," which gets its name from the heat chamber that some printers have. Normally in this option, the temperature of the printing bed, if your printer has a heated one, is the most interesting part. The temperature differs from the material you are using and even from the temperature of the printer environment. Normally, for PLA, a temperature from 120°F to 160°F will work fine. **(Illus. 08.03)**

Comb

This is a nice tool to avoid fine threads of material between the parts of printed objects. Because the plastic is fluid, it has the propensity to drip off the printing nozzle; if it has to go through midair, it sometimes will take thin threads like spider webs with it. To avoid such spider webs, you should avoid driving through midair; "Comb" does

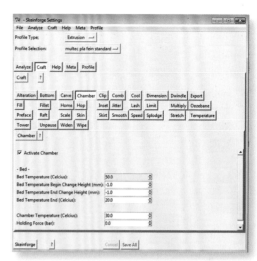

08.03 In "Chamber" you can enter the temperature, for example, of the heated printing bed.

08.05 Many spider webs can be avoided by using "Comb."

this work for you. You should activate this option by selecting the tick mark in this box. The value given by default is fine. **(Illus. 08.04), (Illus. 08.05)**

Dimension

This is one of the most important options to get a good printing result. "Filament

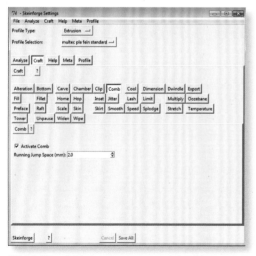

08.04 "Comb" is important to avoid spider webs due to moving the printing head through midair.

Diameter" is self-explanatory, but it is really important. Normally the printing filament is advertised as 3 or 1.75 mm, but the reality is often not so. There are some variations in filament diameters, and often a 3mm filament will be 2.85mm in reality. This will make a great difference to the print quality because, in this case, there will be not enough material available for the print, which will result in a poor surface finish or even a faulty printout. So check (for example with a caliper) the actual diameter of your filament when you get some from a new supplier, or, better still, every time you get a new spool, because it may vary from one production batch to another.

The "Filament Packing Density," e.g. the density of the plastic material in the filament, may also vary, so you may need to change this value as well. Often the default value of 97% will work fine.

In the previous options, I have already mentioned the spider webs that can result from the printing nozzle traveling through midair. You can minimize them by retracting the filament from the printing nozzle when such movements are unavoidable. Shortly before the nozzle travels through midair, the filament can be retracted from the nozzle and, just before it starts printing again, it will be pushed back into the nozzle. These values could be entered here under "Extruder Retraction Speed," "Retraction Distance," and "Restart Extra Distance." The value of "Restart Extra Distance" should normally be a little bit less than "Retraction Distance" so that it does not produce a drop of material in the new printing area. If printing in the retraction area does not work properly, you can experiment with these values.

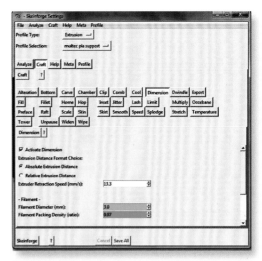

08.06 In "Dimension," you can adjust the diameter of the filament.

It is important that you put a tick mark in the box beneath "Retract within Island" because otherwise these setting will not be used anyway. **(Illus. 08.06)**

Fill

Normally you will not print an empty object, so you have to fill it with material. So "Fill" should always be activated.

Under this option you can set the value for "Extra Shells," which means that not only one massive layer will be put around the outer side of the object, but more than one, which results in a massive shell.

Also very important is the "Infill" option. Here you should enter which pattern the object should be filled with and, even more importantly, which solidity the object should have ("Infill solidity"). Here everything is possible between a totally empty object and a completely filled one. The value is given in decimals, so 0.0 means 0% filling (empty) and

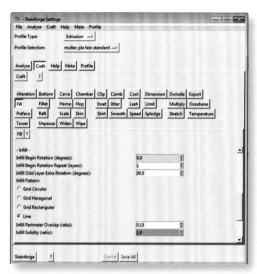

08.07 In "Infill," you can adjust the degree of filling of a printed object.

1.0 means 100% filling (massive). Normally a value between 0.3 and 0.4 will work well so that the objects are stable enough, not too heavy, and don't use too much material. **(Illus. 08.07)**

Raft

Some of the most important values for a successful printing will be entered in the "Raft" option. For some more detailed prints, I will describe this later in "Printing Practice"; it is necessary to make some additional constructions, because otherwise they will not print properly.

For prints that do not have a large bottom area, it may be problematic getting the print to adhere to the printing bed. This means it will be necessary to add one or multiple special layers to make the adhesion area bigger. This is called a "base" or "raft" and would be activated here. If this function is used, an easy to remove extra layer will be printed under the first layer of the part being printed. This will let the part adhere to the printing bed and avoids a separation of the part from the bed during the printing process.

Even more important is the "Support" option, because you cannot print with the filament in midair. For structures that are not formed on other parts of the printed part, you have to build a supporting structure, which will be removed after the printing is completed.

Under the "Support" option, you can enter the important values for this function. "Support Cross Hatch" means that every single layer of the support will be turned by 90° so it is easier to peel off the support material after printing. With "Support Flow Rate over Operating Flow Rate," the quantity of material used for the support can be defined. This means that if it is below 1.0 there is less material used for the support than for the printed object, which also makes the support easier to remove.

The option where support material is printed under "Support Material Choice" is very important. If you choose "None," no support will be printed anyway, "Everywhere" means supports will be inserted under all overhanging parts of the part, whereas "Exterior only" means that support will only be inserted at the overhanging parts on the exterior of the part. This is very useful because overhanging parts in the inner printing area of the part are often underpinned by other parts of the printed part and do not need an extra supporting structure. It is often very complicated, if not impossible, to remove the support structure from the inside of a printed part. **(Illus. 08.08)**

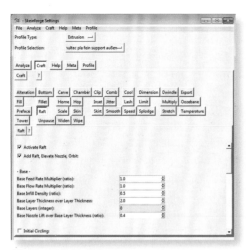

08.08 With "Raft," you can activate the use of a raft for very fragile parts.

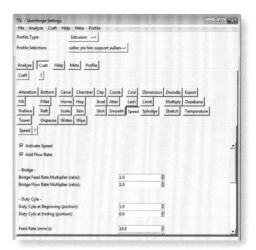

08.10 The "Speed" option opens the field for many interesting experiments.

Skin

"Skin" is a tool to improve the quality of the surface of printed parts. It is very useful but quite difficult to fine tune; you should experiment with different settings to get

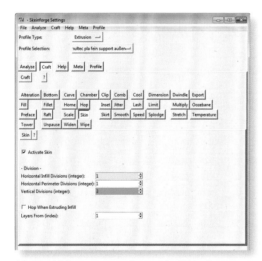

08.09 With some experiments in "Skin," you can improve the surface of your printed parts significantly.

a good result. The online Skeinforge help system has a lot of drawings that help to explain how you can improve your results. **(Illus. 08.09)**

Speed

In the "Speed" option, the velocity of the printing head in the X and Y-axis ("Feed Rate") is entered, and also the speed with which the filament is transported through the extruder ("Flow Rate Setting"). Both affect each other and depend very much on the capabilities of your printer. An important rule of thumb is that if the layers your printer produces are too thick (thicker than you want), the flow rate is too high, and if the layers are too thin, the flow rate is too low. **(Illus. 08.10)**

Temperature

This option is mostly self-explanatory. Different filaments may need different temperatures for a successful printing, and

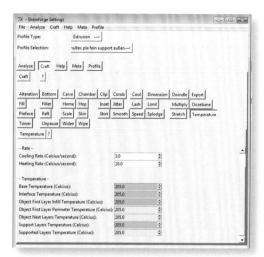

08.11 "Temperature" can be an interesting point for improving the printing quality.

you enter this here. It is also possible to enter different temperatures for different layers, but due to the slow temperature change of the heating element, I do not think that this

usually makes a big difference to the result. **(Illus. 08.11)**

Working with Skeinforge

Saving the profile settings when working with Skeinforge is quite simple. After you click on "Craft," you choose an STL file that you want to process for the printing process. Skeinforge will calculate the g-code for the printing and, after this has finished, will show in a graphic the print path for every single layer. Using this graphic, you can check if everything is okay. For example, you may have forgotten to activate the processing of support information.

The g-code should be saved in the same folder as the original STL file with some information so you can identify it easily. **(Illus. 08.12), (Illus. 08.13)**

You can then load the finished g-code into your driving software and start the print.

Another very interesting slicing program is Cura, which is provided by the Dutch printer

08.12 The calculating process of Skeinforge is quite unspectacular.

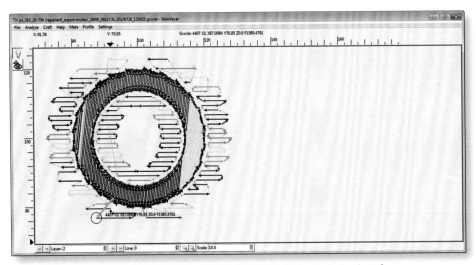

08.13 Here you can check the correctness of the calculated g-code in Skeinforge.

manufacturer Ultimaker. You can find it at http://software.ultimaker.com.

Cura is not only software for slicing, but can also be used for driving the printing process itself. The usage is very simple, with a wizard for the initial setting up that is very helpful, especially when you do not have an Ultimaker printer. **(Illus. 08.14)**, **(Illus. 08.15)**, **(Illus. 08.16)**

The other settings are done from two tabs in the main program. Under "Print config" are the main settings that we have

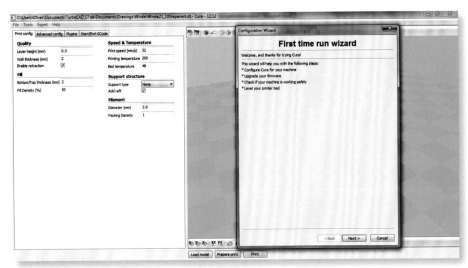

08.14 If you run the wizard in Cura, you can easily enter the needed information.

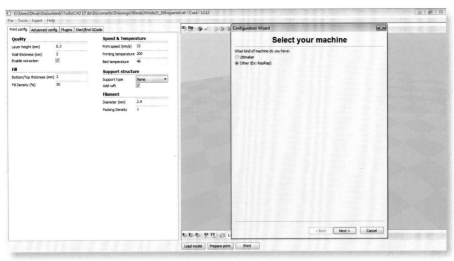

08.15 First you have to choose the machine type you are using.

already seen in Slic3r and Skeinforge. Here you can enter the "Quality," which means the height of the layers, the thickness of the walls, and if the filament should be retracted or not.

"Fill" will determine the thickness of the top and bottom layer and the density of the filling. "Speed & Temperature" is self-explanatory, and we have already seen what the options in "Support structure" are. After you have checked

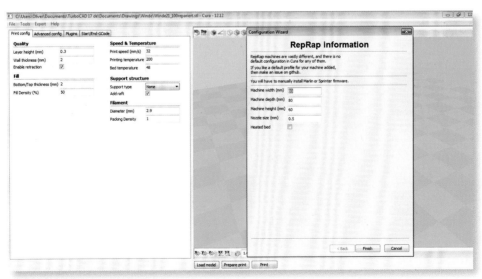

08.16 Then you type in the main data for your printer.

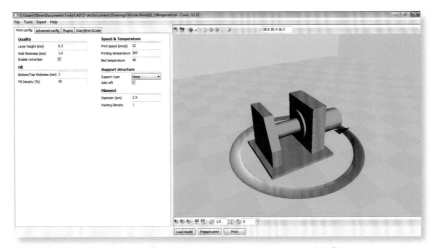

08.17 In Cura, the main configuration parameters are summarized in one screen.

the values for the "Filament" used by you, you can directly generate the g-code. (**Illus. 08.17**)

For more possibilities, especially if the normal settings will not work, it would be helpful to take a look at the tab "Advanced config" and enter some more options like the size of the nozzle, the retraction, and some more settings to improve the quality of the print. (**Illus. 08.18**)

After these settings, if the print was not of the required quality, you will find some "Expert config" under the tab "Expert." There you can modify all the settings similarly to the other programs. (**Illus. 08.19**)

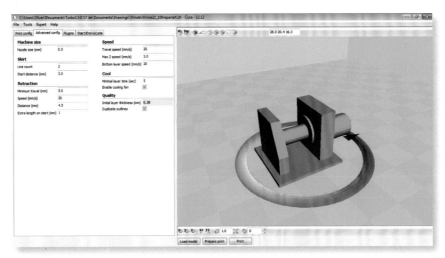

08.18 The different parameters are summarized in another screen.

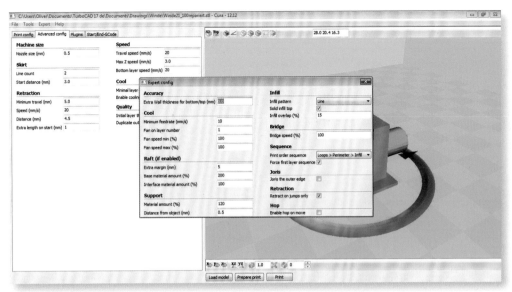

08.19 For more complicated requirements, you can use the expert settings.

You can see that Cura is a little bit more methodical than Skeinforge. First you work with the "Print config," then with "Advanced config," and after this with "Expert config." This makes Cura a little bit easier to understand than the more confusing Skeinforge.

Which slicing program you use depends on your preferences and on the part you want to print. I use Skeinforge, Slic3r, and Cura in parallel. My favorite program is Skeinforge, but some files unfortunately are not sliceable with this; I don't always know why. If Skeinforge won't slice the part, I use Slic3r, which is not so delicate with problematic data, although it does not produce quite so small g-code files. For special uses. I use Cura. It is useful to install a variety of slicing programs and experiment with them.

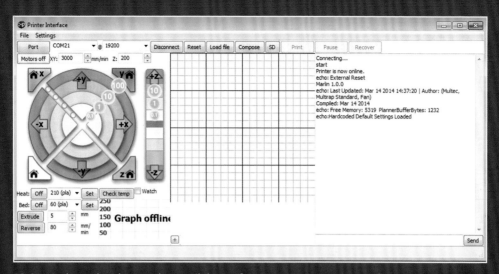

09.01 We have already seen the possibilities of Printrun as an easy-to-use driving program for 3D printers.

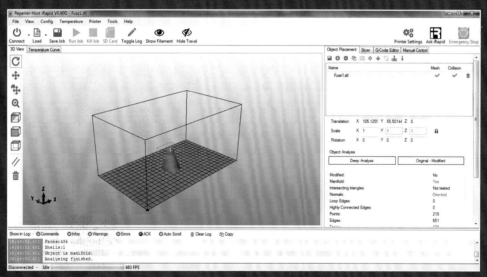

09.02 Repetier Host is often used for driving 3D printers, and you can do a lot more with it as well.

9

Driving software

Your 3D printer not only needs g-code to produce a printed part, but also something that can read this g-code and tell the printer what to do. The software needs to drive the printer's movement, control the heating processes (bed and nozzle), and check and regulate the essential parameters.

Normally your printer will be delivered with driving software. Some slicing programs can also be used for driving the printer. In Chapter 4, "The first print," I introduced a rather basic but very good driving program called Printrun. There are several other driving programs to introduce, and you will see that they are all very similar in use, although some look quite a bit different. **(Illus. 09.01)**

One of the most common driving programs is Repetier Host, an open source program which has really complete functions. It is possible to integrate slicing programs like Slic3r or Skeinforge into this program so you can slice your printing part and start printing straight away. Repetier is a very nice and powerful tool. **(Illus. 09.02)**

Often your printer will come to you with a customized version of Repetier that is already installed with all the information that it needs for printing. If you would like to use Repetier with a printer that was not delivered originally with Repetier, you can enter the data of your printer by clicking on "Printer Settings" on the upper right side of the screen. After this, you enter the data for the connection between the printer and computer, the data for the feed rate and heating, and the shape and dimensions of your printer. **(Illus. 09.03)**, **(Illus. 09.04)**, **(Illus. 09.05)**

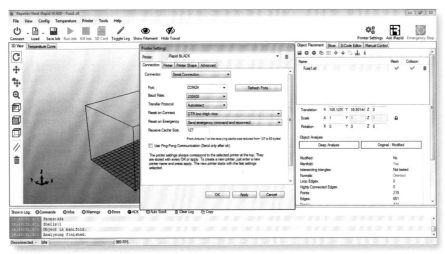

09.03 Normally your printer will come set with default settings for your printer, but you can change any of the settings manually.

After setting the program up, you can work with Repetier, which is quite simple. Connect your printer by clicking "Connect," and you will have full control of the printer. You can (as described for Printrun) heat the extruder and the heated bed and move the axis by hand after you click on "Manual Control" on the right-hand side. (Illus. 09.06)

To print a part, you have to load the data; if you have a suitable g-code program for

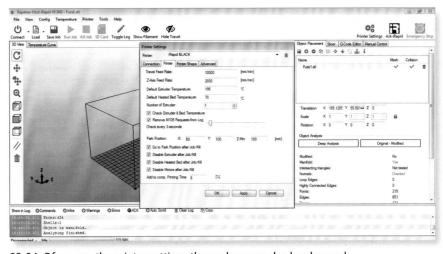

09.04 Of course, the printer settings themselves can also be changed.

09.05 The printing area size can be changed as well.

your printer, you could load this straight into Repetier and start printing. Normally you will have an STL file; the great advantage of Repetier is that it can be integrated with a slicing program. It is also possible to implement more than one slicing program,

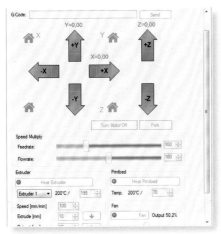

09.06 In Repetier, the printer can be driven manually.

such as Slic3r and Skeinforge. You load the STL file, which will be shown virtually on the Repetier screen, and after that you choose the slicing program you want and click on "Slice with…" The g-code will be generated and you can start the printing process immediately. Alternatively, you can let the printing begin automatically after the g-code is generated. Repetier will control this for you. **(Illus. 09.07), (Illus. 09.08)**

For printers with a stand-alone ability, which means they do not need a connection to your computer during the printing, Repetier allows you to load the g-code into a memory card like an SD-card. **(Illus. 09.09), (Illus. 09.10)**

Another driving program that we already know is Cura by Ultimaker. Cura is not only slicing software but is also a very good driving program for Ultimaker printers and works with nearly every other 3D printer. **(Illus. 09.11)**

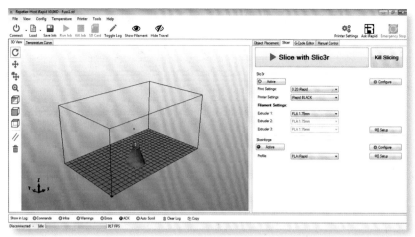

09.07 In Repetier, you can slice the object for the following print. If you have profiles for several slicing programs stored, you can choose between them; in this case the options are Slic3r and Skeinforge.

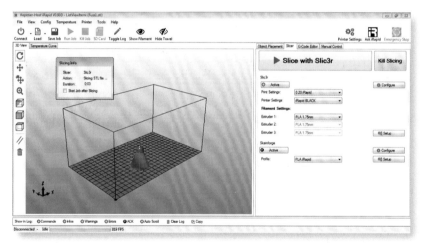

09.08 After clicking on "Slice with...," the slicing process and the calculation of the g-code begins automatically.

Opposite top: 09.08 After clicking on "Slice with...," the slicing process and the calculation of the g-code begin automatically.

Opposite center: 09.09 After the g-code is finished, the printing can be started.

Opposite bottom: 09.10 You can save the g-code (for example, on an SD-card) for printing later on. With some printers, you can print without a connection to your computer.

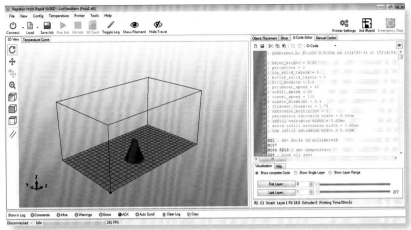

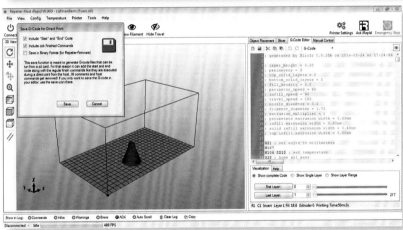

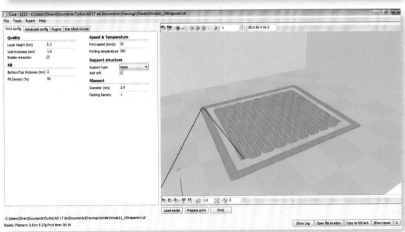

10.01 In this printing, the bridges were too long, so some threads are sagging.

10

Printing practice

3D printing is not like printing a letter on a normal printer. Although some 3D printers are already at a stage of development that make them easier to use than in the past, there are still a lot of special things to learn, like constructing the parts that you want to print, preparing the print and the printing itself, and the finishing process of the printed parts.

With nearly every print that you make you will gain new experience, become more skilled, and get better results.

Next I would like to give you some hints and tips about the specifics of 3D printing with a home FDM printer. Not all of the techniques are applicable if you get your parts printed by a service company like Shapeways or i.materialise, because they use different printing methods that allow different construction techniques and often do not need as much finishing as FDM parts require.

Construction

It is not important which CAD program you are using to construct your printed parts, but with all of them you have to pay attention to some special things during construction, because otherwise it may not be possible to print your part at all or it may be very difficult to finish the part.

Even if it looks straightforward, you should have a very serious look at the printability of your parts. For example, it will be difficult to print very sharp-edged parts because the printer is printing with thin, round threads and so sharp edges will become a little bit rounded off. Of course, it is possible to reach a

10.02 This bridge is short and has printed properly.

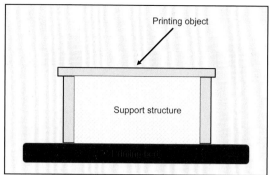

10.03 This object is oriented the wrong way. A lot of support structure will have to be printed.

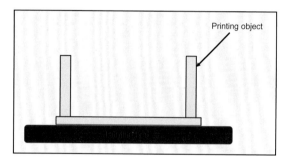

10.04 By simply changing the orientation, no support is needed.

certain sharpness, but do not be disappointed if the objects you are printing seem a little bit smoother than you thought because of your construction.

You should also avoid, if possible, very long bridges in your prints. This means too long a distance over which the material has to be laid in midair; if they are too long, such bridges will have to be supported by material that has to be removed later, often with great effort. If the supports are lying inside the print, it may be difficult or impossible to remove them. Normally, bridges that are up to one centimeter or half inch long are not a problem and will be printed correctly. There is a possibility that the first threads may sag a little bit and may be removed, but this should not be a problem. **(Illus. 10.01)**, **(Illus. 10.02)**

Another simple trick to get good results is the correct orientation of the printing object on the printing bed. Ensure that the object you are printing is connected to the printing bed with the biggest possible surface, even if you have to print the part upside down. For example, if you want to print a table-like part, it will be best to print it oriented so that the table top is lying on the printing bed. Then the table legs will be printed on the underside of the table and no support will be needed at all. If you print the table the "right" way up, it will be necessary to print supports between the legs until the table top can be printed on top of the legs. These supports will have to be removed after the printing. **(Illus. 10.03)**, **(Illus. 10.04)**

It is always a good idea to print a smooth surface, for technical or optical reasons, directly on the printing bed. Printing on its plain surface will make the surface of the printed part very plain and smooth if a raft is not used.

10.05 The part to the left is constructed too finely, making it unprintable. The part on the right has walls that are thick enough to print.

The precision with which you can draw in a CAD program is fascinating. You can draw even the smallest parts with a very high precision, but you might not be able to print them due to the limits of your printer.

So take a critical look at the printing of objects that you have sketched. For example, if you are printing with a thread with a diameter of 0.5 mm, it will, of course, not be possible to print parts that have smaller dimensions than 0.5mm. Some slicing programs will automatically use the minimum possible dimension, but others will make g-code that will not print anything at all. Your printer may sometimes print such structures and sometimes not. This is always a trial and error game. **(Illus. 10.05)**

An important aspect of the printing quality is, of course, the thickness of the printed layers. But the simple formula "the printing result is better if the layers are as thin as possible" is only half of the story. Basically this statement is right, but there are some factors that stop this formula from working. Due to printing thinner layers, the printer has to make more passes over the printed part to build it up, so the printer has to be adjusted very precisely to put the layers on in the correct position. Precision adjustment and axes movement is the basis for a perfect setup of the printed layers.

Another point that you should remember is that using thinner layers will need much more time to build up a part than using thicker layers. You should decide if the part is worth

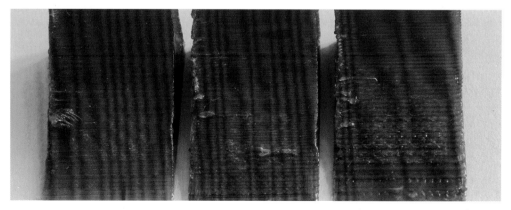

10.06 The same part (a curved surface) as printed with different layer thicknesses: from the left, 0.08 mm, 0.125 mm and 0.25 mm.

the extra time taken to produce it. If the quality of the surface is not that important, it will be better to print thicker layers to get your print done faster. **(Illus. 10.06)**

One point that is a key factor for printing with FDM technology is the use of the support structure. It is clear that you cannot print through thin air. To print parts that have no structure on which they can be built, you have to print a helping structure, a support. A support is, if activated, calculated by the slicing software automatically, so you do not have to construct it yourself. You can normally change some parameters in the slicing software to change aspects of the support (for example, the solidness of it). Often only a minimum support is necessary to support the printed parts above it. The less material that is used, the easier the support can be removed after printing, which means faster finishing and often a better surface on the printed part.

You can also enter at which overhanging angle threshold the support should be printed. Sometimes, until a certain angle is reached, no support is needed; you can figure this out by trial and error. Support may only be needed in the outer regions of the printed object, because the interior has enough bridges to support the printing. Again, this can only be figured out by trial and error.

It is important that you print as little support structure as possible because it needs to be removed after printing, which causes work and takes time. The support structure will nearly always leave some marks on the surface after removal, which will need some work to finish.

Another very important point in the printing practice is the degree of filling

the object is given. The constructed part is normally only a shell with no filling at all. Your slicing program will calculate the degree of filling after you enter the parameter you want. Normally it is possible to choose a filling between 0%, which means the shell will be empty, and 100%, which means the object will be filled completely with material. Normally these values are entered in decimal numbers, so 0.0 means 0%, 1.0 means 100%, and everything in between is possible. Normally a degree of filling of 30-40% is quite enough to make a very stable object. Less filling is possible if the object does not collapse due to insufficient support in the interior.

More filling can be a problem because shrinkage of the cooling material will be higher if the mass of material is rising; this may cause warping of the print. This warping may result in the printed part detaching itself from the heated bed during the printing, resulting in a failed print. A very high degree of filling will also need more material and printing time. **(Illus. 10.07)**

10.07 This part was printed with a filling density of 30%, making it stable enough.

10.08 These printed parts lost contact with the printing bed.

Another very important point for a successful print is the correct adhesion of the printed part to the printing bed. For printing with an FDM-printer, plastic is heated and melted. After leaving the printing nozzle, the plastic will begin to cool down. Material that is hot has a higher volume than when it is cold, so the material will shrink during the cooling process. Even though the degree of shrinking for PLA is smaller than for ABS, the printed object can warp and so may detach itself from the printing bed. If this happens and the part loses contact with the printing bed, the following layers will not be put on the part correctly. This is especially annoying when the print has taken a lot of time and the part peels of shortly before the print is finished. Luckily, there are a few ways to prevent this problem. **(Illus. 10.08)**

One way is by using a heated printing bed that is very useful supplementary equipment for most 3D printers. Because of the lower temperature difference between the molten printing material and the hot printing bed, shrinking of the plastic is not as high as it would be if printed onto a cold bed. The effect is that the printed part sticks to the printing bed better than a bed without heating.

The printing bed is usually heated by a heating pad that consists of glass or silicone. The heating pad is glued or screwed under the metal printing bed or, in some printers, the print is done straight onto the glass heating pad. The temperature that is needed

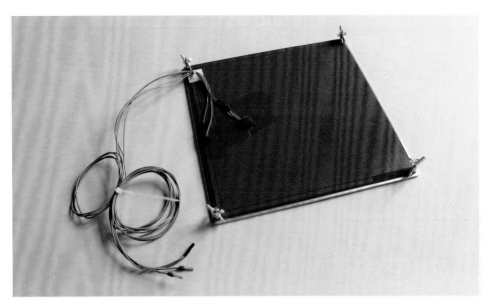

10.09 A view of the underside of a printing bed. The red part is the heating pad.

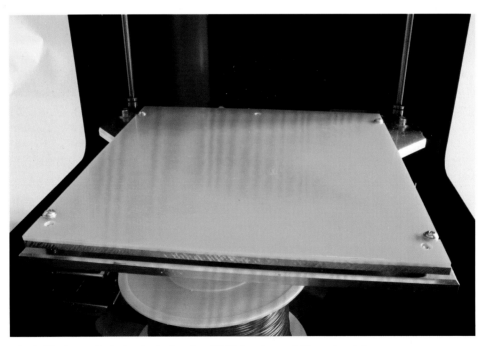

10.10 The heated printing bed of an Easy3DMaker (www.3dfactories.de).

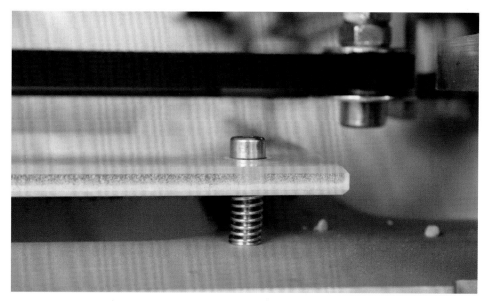

10.11 This glass printer bed is also heated.

depends on the material being printed. For PLA, you need a temperature of 120-140°F for the material to adhere to the bed. ABS needs a much higher temperature of 180°F or even more than 210°F.

The driving of the heating bed will normally be regulated by the electronics of your printer and the printing software. Here you can simply enter the required temperature and the electronics will control it. A nice feature is the ability to turn off the heated bed automatically after finishing the print. If the print is finished and you are not in the room, or you are printing overnight, you will save electricity. Also, the printed part can be removed from the heated bed easily when it has cooled down, the little bit of shrinking being advantageous. **(Illus. 10.09), (Illus. 10.10), (Illus. 10.11)**

Another method for getting good adhesion of the printed part to the printing bed is by using certain design features during the construction of your parts. One way is, for example, to construct the parts in such a way that they have maximum contact with the heated bed. You remember the example of the orientation of the table? If this is not possible and the contact area is small, you should use a so-called "raft." A raft is a helping structure that increases the contact area of the object. It is printed like a thin support under the first layers of the object.

Printing a raft can be chosen in the slicing program and the raft will be added automatically. In most programs you can enter parameters for the raft that will optimize it for your printing. Rafts increase the contact of your part to the printing bed; just peel off the raft after the print is finished. **(Illus. 10.12)**

To make adhesion of the printing part secure, the surface of the printing bed is also important. Printing beds are often made

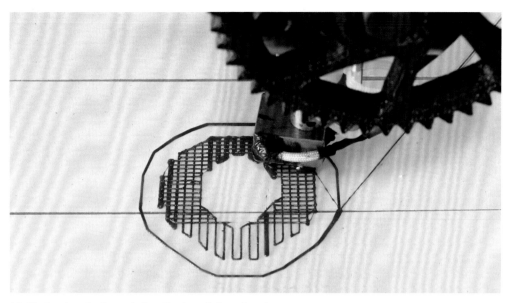

10.12 A printed raft can help adhesion of the printed part.

from metal (mostly aluminum) or glass. Due to their smoothness, a printed part will not adhere very well to the bed and needs some kind of interface between the bed and the printed part. One of the materials to aid adhesion, even without a heated printing bed, is double-sided adhesive tape like the kind you use for laying down carpet. This will help the adhesion, but there may be a problem releasing the part because the adhesive tape will adhere to the plastic, causing the part to break when removing it. Also, some debris from the adhesive may remain on the underside of the printed part, which spoils the look of the part. Double-sided tape is a compromise and is not really suitable for good prints.

Even if your printer has a heated bed, you need to make the bed "stickier." Adhesive tape can be used and crepe tape or the special tape that is used if you are plastering walls is a possibility, depending on your printer and the material you use. Always test what kind of tape gives the best results. **(Illus. 10.13)**, **(Illus. 10.14)**

Another special tape that you can use is so-called Kapton tape; this makes a printed part adhere very well. Kapton tape has a very high temperature tolerance, so it also works for ABS, which needs a much higher heated bed temperature, as we learned earlier. So, for ABS, Kapton tape is necessary, but it also works well for PLA.

Some printers increase adhesion by using glue. There are several "magic cures" that are used for this. They range from thinned wood glue up to very special, often dubious, glues that are not recommended. If you use such glues, be sure they are suitable for your heated bed, or damage of the heated bed or even the printer itself may result. If you are unsure, ask your printer manufacturer.

10.13 Adhesive tape is often useful for good adhesion of the printed part to the bed.

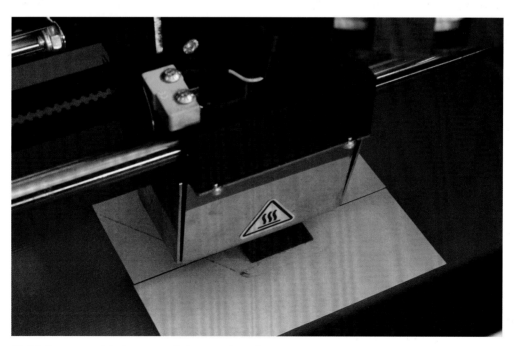

10.14 Even paper labels may work, but test them first.

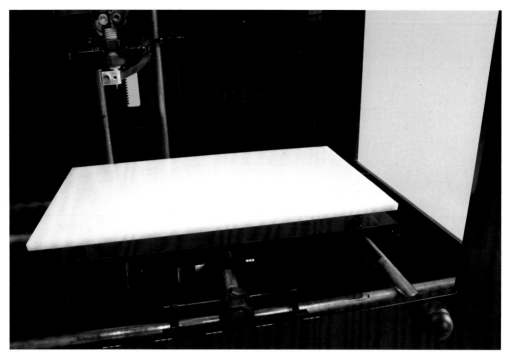

10.15 Some printers, like this iRapid BLACK (www.irapid.de), do not need a heated bed at all.

Printing beds made out of plastic do not usually need a heated bed and allow the part to adhere easily. The disadvantage is that plastic beds often wear out quickly and have to be replaced regularly. They are more sensitive and not so resistant to mechanical stress as printing beds made out of harder materials. However, they are a good option for an unheated bed without using adhesive tape. **(Illus. 10.15)**

Even though we are not baking a cake, preheating is important to get a fast printing start. Both the bed and extruder need time to reach the right temperature. If you want to start your print fast, you should preheat both, especially as a lot of programs will only start printing when the right temperature has been reached.

One thing you should know is that preheating can cause a problem, especially if you let the printer stay heated in an inactive state. As the printing nozzle is open at the bottom, the liquid plastic will drop out of the nozzle. If you do not fill the extruder with material before you start the print, there will be some holes or other imperfect parts in the first layer, so before you print, fill the extruder with material.

Before you start your print, you should check some points to ensure you get good results. The connection between the printer and computer should be working and the g-code should have been prepared for printing. You should also check the following.

1. Is the printing temperature correct? The printer will only start if the right temperature has been reached. The printing software does not know if you have chosen the right temperature. If you print with PLA and ABS and change between the two materials and printing temperatures, make sure you choose the correct temperature. If ABS is printed with the lower PLA temperature or PLA with the higher ABS temperature, both will result in poor quality prints, if they work at all.

2. Is the right filament loaded into the printer? It could easily happen that the wrong plastic, wrong color, or wrong thickness of filament is in the printer. Check that the filament is correct before starting to print. Also check if there is enough filament left in the printer. There is nothing more annoying than not having enough filament to finish a print, especially when the print will take many hours to complete.

3. Is the printer free to move in all axes? It could easily happen that an object (for example, a tool or a printed part) is lying on one axis and will stop the printer from moving. The result will be that the printer stops, the print will not be done, and the printer or components could be overloaded, damaged, or even destroyed.

4. Is the printing bed clean and prepared? As seen previously, the printing bed has to be prepared in the right way before printing. Remains of any previous print should be removed, adhesive tapes may have to be replaced if damaged, and glues have to be put on the bed.

If all these points are okay, you can start the print and be relatively sure that you will soon be holding the printed part in your hands.

One other thing should be checked shortly before the print to see if everything is working properly. Most slicing programs have got an option called "brim." It means an extra loop of extruded material is laid around the print before the first layer of the printed part is done. The reason for this extra frame around the print is that, during this procedure, the printing nozzle is filled with material, so the print starts perfectly. This is the time during which you should observe the printer very carefully. If the brim is not laid correctly after the first few centimeters or inch or two, there could be some problems during the print; either the temperature may be incorrect or the filament movement is not working right. If the brim starts fine and then becomes holey, it may be better to stop the printing and fill the nozzle with material by hand. **(Illus. 10.16)**

An important point during printing is changing the filament, whether the material or color. Both should be done very carefully.

If you change between PLA and ABS, you should heat the nozzle to the higher temperature, in this case the temperature of ABS, before the change. After the temperature is reached, change the filament to the new one and extrude a few centimeters or inches of the material to remove the old material completely.

The same procedure should be done during the change from one color to another,

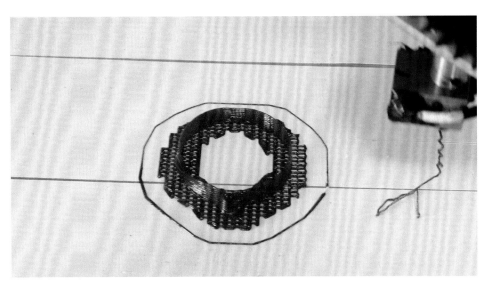

10.16 The brim is a loop around the printed object.

especially if you change from a dark color to a lighter one. You should extrude the filament until all the darker material in the nozzle is removed. Otherwise, darker material will appear in the lighter color, which may result in unattractive dark spots in the printed part. **(Illus. 10.17)**

Not all prints leave the printer in the way you need. In some cases the rough surface will not be suitable; some will be needed in

10.17 When changing the filament color, you have to manually extrude the material until the new color comes out of the nozzle clean.

a different color, or a combination of colors, that are not available or not possible. If needed, prints can be finished with normal craftsmanship. The rough surface can be sanded with normal abrasive paper and the color can be changed by painting.

Nearly every finishing technique can be used for PLA and ABS; sanding (wet or dry), milling, and drilling is possible. Of course, for every technique, care has to be taken, as thermoplastics have a propensity to melt and become greasy.

Even gluing 3D prints is possible. The best glue for PLA is cyanoacrylate, often called superglue. ABS is soluble with acetone and can be glued with acetone-based glues. PLA cannot be glued with acetone.

Almost any paint can be used for lacquering printed parts. If you are not sure if the material is affected by the solvent or the paint being used, test it on a sample of the plastic first. **(Illus. 10.18)**, **(Illus. 10.19)**

Like every machine tool, a 3D printer needs a bit of maintenance. A good working printer should be kept clean with mild detergents that can be used for electronic and sensitive mechanical devices. If you use hard detergents, it may result in damage to the printer and invalidation of the warranty.

Protect the printer from dust by covering it whenever it is not in use with a cover like a blanket or similar; this will eliminate a lot of cleaning need.

Check the mechanical components of the printer regularly. Due to the fast movement of the printer axes, there is a possibility of screws or other connecting parts coming loose. This may result in an unacceptable slackness in the movements, which will have an effect on the quality of the print or

10.18 You can sand the printed object to a smooth surface.

10.19 Printed parts can be painted with normal paint.

even make printing completely impossible. Check if all connections are tight and if the clearance of the moving parts is good: not too slack, but not too tight either.

The frame assembly should also be checked regularly, because a moving frame can transfer

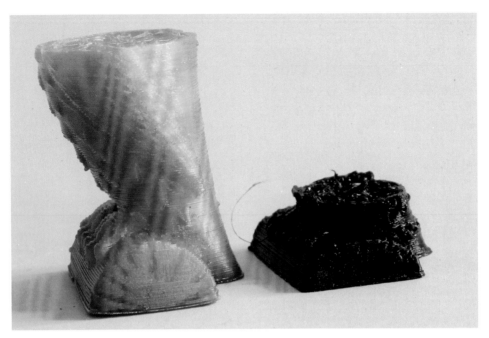

10.20 On the part to the right, the Z-axis did not move correctly because of a loose screw. The red part should have had the same height as the orange one.

vibration to the moving parts, resulting in poor-quality prints.

You should check that any moving parts, such as toothed belt pulleys, are fixed correctly. If they are loose, they may slip on their shafts and result in poor print movements; the result will be a bad print. A damaged toothed belt will also make a good print impossible, so check out the toothed belts regularly.

The printing bed should be checked from time to time. Are there too many marks from contact with the nozzle or from the removal of printed parts? Has the bed been bent during the removal of parts? If so, you should mount a new printing bed, although this will only be necessary on rare occasions. **(Illus. 10.20)**

The moving parts, such as threaded spindles and bearings, should be lubricated occasionally. Use very fine silicone oil where possible. Make sure that no lubricants get in contact with the printing bed, because this will stop the printed parts from adhering to the bed. **(Illus. 10.21)**

An important part of your printer is the transportation system for the filament. Most printers transport the filament using two rollers; one has small teeth and transports the material into the extruder, whereas the other one is smooth and presses the filament onto the toothed roller. Normally the smooth roller is tensioned by adjustable springs. Using this system, the alignment of the filament transportation can be optimized for best results. You should adjust the system

10.21 From time to time you should oil the bearings in your printer.

10.23 The filament transportation system of the Creatr made by Leapfrog.

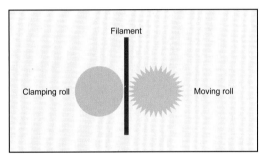

10.22 A schematic view of the filament transportation system.

10.24 Insufficient material was transported to make this print.

so that the filament can be easily moved but the pressure on the filament is not too high. If the pressure of the rollers is insufficient, the filament will not be transported into the extruder. If the pressure is too high, the toothed roller will abrade the surface of the filament and result in problems like the filament jamming. Adjust both rollers carefully and check them from time to time. If there is abrasion seen at the toothed roller, or you hear clatter from the stepper motor, the tension of the filament transportation is too high, so you should loosen it a little bit. If the printed part has an incorrect shape, you should check if the filament is being transported correctly, because there might not be enough material being brought into the nozzle. (Illus. 10.22), (Illus. 10.23), (Illus. 10.24)

11.01 With a little construction work, even multicolor prints are possible without
a multi-extruder. *(see page 121)*

What can I print?

Even if I am tempted to say "everything," this will, of course, be wrong. Some recent media coverage gives the impression that 3D printing will be the solution for all the problems of the world, which it is not. It is a new manufacturing technique that opens a lot of possibilities that were not reachable until now.

The main restriction when using home FDM printing is the usable material, which is always a thermoplastic. This will restrict the things you can print to parts that do not need resistance to high temperatures or resistance to special chemicals like acetone.

For example, it is not possible to print a coffee cup out of PLA, because PLA will soften at temperatures higher than 140°F. Once, I tried to clean a printed cookie cutter made of PLA in the dishwasher; after washing, it had a nice abstract look, but it looked nothing like the cookie cutter it was originally...

The mechanical resistance of printed parts made out of thermoplastics with the FDM technique is limited. However, you will be surprised at how robust FDM printed parts can be when the construction is done in the right way. If the plastic itself is not stable enough, it is possible to strengthen a printed part with metal structures such as rods. A nice feature of 3D printing is that you can print exactly-sized holes in the parts where strengthening structures can be added after printing.

It is even possible to print a part with a totally enclosed strengthening structure. You print the first half of your part, inside of which you have constructed a kind of slot to fit a strengthening part in. When half of the slot is printed, pause the print and insert the strengthening structure, then restart the printing. The slot will be closed and you will not see any part of the strengthening structure.

If the stability of the thermoplastics is sufficient, there are only minor limits to the things you can print. The main challenge is to construct the part you want to print in a way that makes it possible to print it. I have already shown the example of orienting a table to get a quality print. We have seen that the orientation of the print is important, but, of course, it is not the only important thing. After you have been printing with your machine for some time, you will know what is possible with it and what is not. This is the same principle as other machines, which have their own strengths and weaknesses. As fine and fragile as parts can be, they should be printed correctly. If you are constructing a spare part or a special helper, you will discover what is possible and what is not. Trial and error is the best way to find out these things, because every different printer and even the different filaments will react in different ways. Sometimes, if a print does not work, it is necessary to change some of the parameters of the printing, e.g. the temperature or the speed, to make it work perfectly. Constructing and printing is often a meticulous fine tuning of all the parameters over which you have control.

For some special uses there are special materials, which are often based on PLA, that have special characteristics. For example, it is possible to print your own "rubber" stamp easily on your 3D printer using a special kind of soft PLA that seems to be a little bit like rubber. This soft PLA is also useful for special suspension parts, etc. If you find a purpose for which you would need a printed part, but you find your normal material is not suitable, look for another type of printing filament, because a lot of new materials are being developed regularly. Maybe you will find a material that will solve all your problems.

One limiting point is the technical printability of an object. If your printer is able to print the part to the required resolution, the part will be printable. Another question: is it reasonable to print a part and produce it combined with other techniques? For example, if it is easier to make an object out of another material, would it be best to make it using other methods and combine it with 3D-printed parts? In some cases, the removing of the support structure will be so complex and time-consuming that it will not be reasonable to use 3D printing for making the object and another technique may be better.

A special feature is multicolor printing, which is now possible with some printers or can be fitted as an add-on. For multicolor printing you need a multi-extruder that can print with more than one filament. This is a very nice feature, but it is not very cheap. Some multicolor parts can be made with a single extruder. For example, you can print every color in a special layer or print different colored parts like a kit and assemble them later. **(Illus. 11.01), (Illus. 11.02), (Illus. 11.03), (Illus. 11.04)**

11.01 With a little construction work, even multicolor prints are possible without a multi-extruder. After printing the white base, the blue parts are printed...

11.02 ...followed by the red parts.

11.03 The finished United Kingdom flag. (Thingiverse, thing:20434 User Alzibiff)

11.04 Many items can be printed on a 3D printer.

12.01 These parts are for building models. (Some of them are from Thingiverse users erdinger, NewtonRob, odie_wan).

Uses for a 3D printer

Now that you have a 3D printer and are able to print with it, what should you print? Maybe this question sounds funny, but it won't in a moment. I know more than one person that has built (or bought) a 3D printer because they were fascinated by the possibilities of this revolutionary process. Then, after everything was working, they asked themselves what to do with the printer. Continuously printing the same decorative figures, vases, or abstract parts may be technically thrilling, but you (and your partner!) proabably want to find a use for this expensive equipment.

I would like to give you some examples for applying 3D printing at home.

Model building and model railways

If you are building model ships, planes, tanks, or other models, or if you build model railways, a 3D printer will allow you to make special parts for your hobby. An advantage is that you can draw the parts that you need for your model to the original measurements and then scale it down to the size you need. Maybe, for example, you need winches for a ship model or a gas can for your model truck. You can download a 3D CAD file of the original part and scale it down for your own use.

You can also draw a part once and scale it for the project you need. For example, if you have printed a part for a ship in 1:48 scale and need the same part for another model in 1:96 scale, you only have to scale it down and

12.02 Here is a printed street part for a model railway.

print it again and the work is easily done. 3D printing is a big advantage, especially for parts that you need more than once. You need 16 lifeboats for a liner? Draw it once and print it as often as you need to!

Although making parts for models using 3D printing is very useful, the printing of a precisely fitting fixture is a great help for some technical equipment like attachments for servos or accumulators.

For model railways, it is possible to draw your own house in a simple CAD program, scale it down, and print it. So your own house will be on your model railway and, if you are ambitious enough, your entire street or town could be printed in a small scale. **(Illus. 12.01), (Illus. 12.02)**

Home and garden

A great field of activity for construction and printing is the area of your home and garden, giving you a good opportunity to show your partner that the expensive printer you purchased was a good investment…

So if you need something in the garden, perhaps a tool for sowing seeds or a special part for fixing a problem in your home, your 3D printer could be very useful and open the way to an exclusive solution for many problems.

Simply click through www.thingiverse.com to get a lot of ideas and see things that you need; most of the time, you will not even have realized that you needed them. **(Illus. 12.03), (Illus. 12.04)**

Top: 12.03 This garden hose holder was constructed to fit on a fence.

Center: 12.04 We have already seen this tablet holder for the kitchen.

Bottom: 12.05 A little toy helicopter (Thingiverse, thing:5225, user t1t4) is only one of the unlimited possibilities for printing toys.

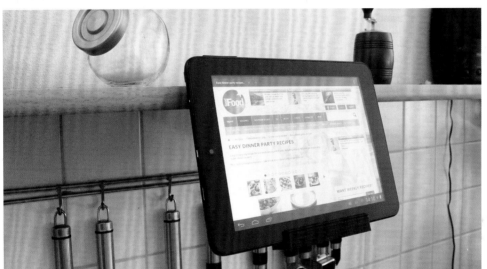

Toys

I want to give you some advice: do not show your children Thingiverse! Your printer will not have one second to cool down if they discover it. There are lots of things there that kids "need."

Seriously, there are really nice things that you can print for children, some of which are really useful. Maybe the construction of a part that your children need would be a good practice for using your construction software. **(Illus. 12.05)**

Decoration and jewelry

Do you need a suitable decorative object for your new home or an extravagant piece of jewelry for the next party? Maybe 3D printing could be the answer to that. Lots of designers (professionals and hobbyists) are publishing files for such things on the internet for free (or paid) download. You can print these parts on your own.

The special characteristic of 3D printing makes it possible to produce things that can't be produced with normal techniques. The design possibilities are almost unlimited.

Strictly geometric or bionic shapes are as possible as very abstract shapes. Often there is a question if such things are possible to print, and, mostly, they are. **(Illus. 12.06)**, **(Illus. 12.07)**, **(Illus. 12.08)**

12.07 A 3D-printed decorative part. *(Picture: Aroja)*

12.06 Some parts for jewelry from Thingiverse: a ring from user allenZ, two types of bracelet by Pixil3D, and a necklace (printed in one part) by user Belfry.

12.08 A little Buddha printed in 3D, *(Picture: Aroja)*

12.09 The tow coupling of this toy tractor broke...

12.10 ...so a new one was printed...

12.11 ...and it fits perfectly.

Spare parts

One thing that is a possible for 3D printing is the production of spare parts. Some big companies are already considering if it would be possible to give users a file of such a part so that they could print it out on their own printers rather than having a big warehouse full of spare parts for everything.

Even if this is a dream for the future, today, spare parts are a big possibility for you if you have a 3D printer, especially for things that are rather old and for which you can no longer buy spare parts. The battery cover for an old remote control or a knob for that radio from your grandfather can be printed with no problems. As long as you can draw the spare part, you can print it out; the problem is solved, making the old item useable again.

If a part for an expensive toy is broken, a self-constructed spare part could be your rescuer from an expensive situation.

The construction of a part may take a little time and need a lot of thinking, but the result will make you proud and give you another good argument as to why you purchased

12.12 Many broken parts can be replaced with printed ones.

your 3D printer. **(Illus. 12.09)**, **(Illus. 12.10)**, **(Illus. 12.11)**, **(Illus. 12.12)**

Personalized and customized objects

Another use for 3D printing is the production of very small quantities of personalized or customized parts. A ring with the name of your husband or wife is no problem. Paperclips with the name of your own company can be printed in seconds. Special cookie cutters with your favorite sport icon or the logo of your sports club are easily made.

This and a lot more is possible with modern 3D printing technology. There are lots of customizable things in Thingiverse, but with a little thinking, it is now possible to make such things on your own.

One fun thing is that the homepage of cookiecaster.com makes it possible, with a few clicks and an easy-to-use technique, to produce cookie cutters that are unique to you so that you can surprise your loved ones with cookies that no one else can make. **(Illus. 12.13)**

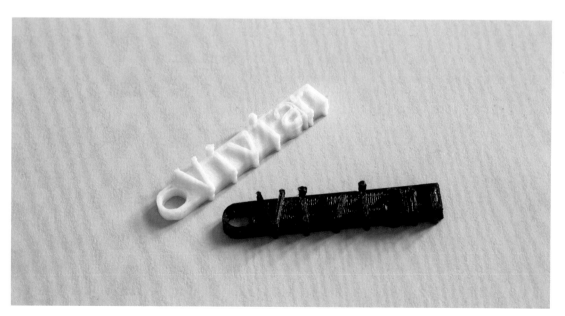

12.13 Personalized objects are a big possibility for 3D printing. Here is a personalized key tag, based on a Thingiverse template by user allenZ.

13

Service companies

To have your own 3D printer (normally working with FDM technology) is not only for printing things. Working with such a printer is also a kind of hobby that gets you into a new technology that will probably have a big influence on our future. Working with a 3D printer will take some time and will not always be successful.

Sometimes, due to economic reasons, owning your own 3D printer may not be possible because the uses of such a printer are limited, the materials are restricted, and the costs are high. If you decide that your own 3D printer is not the right thing for you, this need not be the end of your journey into 3D printing. There are a couple of 3D printing service companies that you can use for producing your prints.

Such companies have several advantages over home printer. They use machines every home user could only dream of, so the prints are often much better quality than a home 3D printer would ever be able to reach, especially because they can use techniques (like SLS and STL) that for the home user are not economic to use at the present time. Due to this, companies with high-end printers can also use materials not suitable for the home printer. Metals like steel, brass, titanium, and even gold and silver are useable, as are special plastics or ceramics. So not only can you order things for limited usage, but you can order specially-designed tableware or valuable jewelry.

Ordering from a service company is very simple. You need an STL file of the thing you want to print. This file is uploaded to the

Above left: 13.01 Due to the different machines service providers can use, you can order self-designed objects like vases made out of ceramics. *(Picture: Shapeways)*

Above right: 13.02 You can make coffee cups too. *(Picture: Shapeways)*

Left: 13.03 Self-designed jewelry, printed in silver or gold, is possible from service providers like Shapeways. *(Picture: Shapeways)*

Below left: 13.04 You can get specialist jewelry made by service providers such as Shapeways. *(Picture: Shapeways)*

server of the service company, the material you want is chosen, and you will be given the price for the print. During this process there will be a check to see if the file is printable. After this, you can decide if you want to proceed and approve the quotation. When you have paid, the object will be printed and shipped to your home.

The number of service companies offering their services is already large, and more are springing up regularly. Some big players like UPS and Staples, for example, are looking for their piece of the cake and thinking over how to offer such a service. Most of these service companies can be found on the internet. Beside the big companies like Shapeways (www.shapeways.com), i.materialise (i.materialise.com), and Sculpteo (www.sculpteo.com), there are lots of smaller companies. Some may be local to your home and can offer you a good range of 3D printing services. The advantage of nearby firms is that problems and questions can be cleared up quickly. **(Illus. 13.01), (Illus. 13.02), (Illus. 13.03), (Illus. 13.04)**

14

Literature and the internet

Like every new and fast-developing technology, including 3D printing, there is a fast-growing list of books and a tremendous number of websites in existence. I will give you some suggestions for books and homepages. This list is, of course, not intended to be exhaustive.

Literature
Chris Anderson:
Makers: The New Industrial Revolution
ISBN 9780307720955

Christopher Barnatt:
3D Printing: The Next Industrial Revolution
ISBN 9781484181768

Carla Diana:
LEO the Maker Prince: Journeys in 3D Printing
ISBN 9781457183140

Brian Evans:
Practical 3D Printers: The Science and Art of 3D Printing
ISBN 9781430243922

Anna Kaziunas France:
Make: 3D Printing: The Essential Guide to 3D Printers
ISBN 9781457182938

Mark Hatch:
The Maker Movement Manifesto: Rules for Innovation in the New World of Crafters, Hackers, and Tinkerers
ISBN 9780071821124

Kalani Kirk Hausman/Richard Horne:
3D Printing For Dummies
ISBN 9781118660751

James Floyd Kelly:
**3D Printing: Build Your Own 3D Printer
and Print Your Own 3D Objects**
ISBN 9780789752352

Hod Lipson/Melba Kurman:
**Fabricated: The New World of
3D Printing**
ISBN 9781118350638

Bre Pettis/Anna Kaziunas France/
Jay Shergill:
Getting Started with MakerBot
ISBN 9781449338657

Internet

www.3ders.org

www.3dprintingnews.co.uk

www.fabathome.org

www.inside3dprinting.com

www.myminifactory.com

www.rapidprototypinghomepage.com

www.reprap.org

www.thre3d.com

www.wired.com

INDEX